Endosymbiotic Theories of Organelles Revisited

Naoki Sato

Endosymbiotic Theories of Organelles Revisited

Retrospects and Prospects

Springer

Naoki Sato
Department of Life Sciences
University of Tokyo
Tokyo, Japan

ISBN 978-981-15-1160-8 ISBN 978-981-15-1161-5 (eBook)
https://doi.org/10.1007/978-981-15-1161-5

This Springer imprint is published by the registered company Springer Nature Singapore Pte Ltd.
The registered company address is: 152 Beach Road, #21-01/04 Gateway East, Singapore 189721, Singapore

Preface

The notion "endosymbiotic origin of organelles" is considered an accepted theory in cell biology that explains the origin of mitochondria in eukaryotic cells as well as the origin of chloroplasts in plant cells. Currently, it is explained in high-school textbooks as if it is an established biological fact. This is a point of discussion in this book. I found many nonspecialists having a serious misunderstanding that cyanobacterial cells are living within the plant cells. Evidently, cyanobacteria and chloroplasts are different. If these two entities were to be mixed up, the basis of cell biology would be collapsed. This is a critical point.

How were endosymbiotic theories born? What was the evidence for the theories? How was "the endosymbiotic theory" established as part of the entire cell biology? What are the proofs for the theory in the current context of biology? Is there any room for doubt in the theory? I am trying to answer these questions one by one. This work is a composite of history of science and presentation of recent studies.

When I was a student in the Faculty of Science, University of Tokyo, in the early 1970s, I attended cell biology lectures of Professors Shichiro Sato and Satoru Murakami, who insisted on the importance of the novel, endosymbiotic theory. This theory became largely accepted in the mid-1970s, when I started to work in the laboratory of Professor Norio Murata, my supervisor at that time. On the other hand, some people, notably Professor Hakobu Nakamura (Kohnan University), strongly opposed the endosymbiotic theory (Nakamura 1987J). One of my first studies involved lipid analysis of cyanobacteria that were then called blue-green algae. At that time, cyanobacteria and chloroplasts were considered to share many biochemical processes, and people believed that what was true in cyanobacteria was also true in chloroplasts. In fact, the two were found similar in ultrastructure: they were surrounded by two layers of membranes, and the photochemical reaction proceeded in the internal membranes called thylakoid membranes. Galactolipids were found as major components in the chloroplast membranes and cyanobacteria. Both cyanobacteria and chloroplasts perform oxygen-evolving photosynthesis using chlorophyll. The presence of chloroplast DNA was already known, and DNA was also found in free-living cyanobacteria. In this way, cyanobacteria were considered free-living forms of chloroplasts. Cyanobacteria and chloroplasts were discussed in identical sessions in the meetings of plant physiology.

With the advances in biology, the notion "endosymbiotic origin of organelles" became established. Recombinant DNA technology that was developed in the 1970s was rapidly applied to chloroplasts to establish the new research field "chloroplast molecular biology." In 1986, two Japanese groups succeeded in determining the DNA sequences of two chloroplast DNAs. Chloroplasts then became the top runner of plant molecular biology and plant genomics. DNA sequences of various cyanobacterial genomes were also reported. The similarity of chloroplast genomes and cyanobacterial genomes was thus established as an undoubted fact. Meanwhile, the notion of secondary endosymbiosis was invented to smartly explain the origin of the plastids of diatoms and brown algae: they are red algal plastids introduced by secondary endosymbiosis. This hypothesis gave an explanation of the differences in the number of membranes that surround the plastids of different algae. At the same time, primary endosymbiosis became a prerequisite for the secondary endosymbiosis.

During the 40 years of my researches, I have been using plants, algae, and cyanobacteria as research materials. My studies were especially focused on the relationship between cyanobacteria and plastids (chloroplasts). This was not intended at the beginning of my research life, but was its results. In other words, I always guided our research from the viewpoint of comparison of cyanobacteria and plastids. As a result, I recently began to have questions about the similarity of cyanobacteria and plastids (chloroplasts), which I had accepted as an established fact when I started my studies.

The year 2017 was the 50th anniversary of the famous initial article of Lynn Margulis on endosymbiotic theory in the *Journal of Theoretical Biology*. A number of articles related to the topic were published in a special issue of the journal in 2017. Many of them admired the works of L. Margulis, but I have various questions about the theory: are chloroplasts and cyanobacteria really similar? Is the apparent similarity of chloroplasts and cyanobacteria a result of functional restrictions imposed by the oxygenic photosynthesis? Then, what is the meaning of endosymbiotic origin of chloroplasts? What is the difference between endosymbiosis and massive lateral gene transfers? Are the visual images of endosymbiotic origin of chloroplasts presented in textbooks valid? To answer these questions, I have scrutinized various articles and books of L. Margulis. I have also read the old article of C. Mereschkowsky in German. The questions that I am trying to answer might perhaps date from the whole history of endosymbiotic theories since the beginning of the twentieth century. That was the reason why I planned to write this book, first in Japanese and now in English. Animating history of biology, which is otherwise forgotten, in the context of current research is my approach leading the narratives in this book. A list of references including many important, though not complete, articles and books is provided at the end of the book to help readers to obtain further details.

Tokyo, Japan Naoki Sato

Notes for the English Version

After publication of the Japanese version in June 2018, some readers suggested me to publish it in English to reach a larger audience. Major articles and books onendo-symbiotic theories have been published in English, but a comprehensive work on the endosymbiotic origin of organelles including both history and current status of knowledge has not been published. As described in Chap. 1, two important books are already available: Jan Sapp's *Symbiotic Associations* (Sapp 1994) and John Archibald's *One Plus One Equals One* (Archibald 2014). The former described historical notions of symbiosis and endosymbiosis culminating in establishment of the endosymbiotic theory of organelles by Lynn Margulis. A more recent historical overview ("Too fantastic for polite society") was given by the same author in a concise form (Sapp 2012) in the book *Lynn Margulis* edited by Lynn's son, Dorion Sagan, in memory of Lynn Margulis (Sagan 2012). The book by Archibald described many anecdotes on the life of Lynn Margulis in relation to the researchers in Dalhousie University, an important center for endosymbiosis studies in Halifax, Canada. Here I try to provide different perspectives (1) by shedding new light on the long tradition of Russian and German studies on symbiosis and symbiogenesis that have been hidden under the sole name of Lynn Margulis; (2) by analyzing the initial publications of Lynn Margulis to focus on her ambitions of establishing the origin of eukaryotic cells; (3) by explaining the reason and background for the acceptance or establishment of the endosymbiotic theory; and (4) by presenting various questions about the currently prevailing, visual images of endosymbiotic origin of chloroplasts. The focus is placed mainly on the chloroplasts or, more generally, plastids. In my understanding, the endosymbiotic origin of mitochondria is currently discussed or questioned, mainly due to the discovery of mitochondria-related anaerobic organelles devoid of DNA. Historically, there were always ambiguous arguments on the origin of mitochondria (see Chap. 3), and reasonably acceptable phylogenetic evidence was presented in the late 1990s. The discussion is still continuing. By contrast, endosymbiotic origin of chloroplasts seems quite established. In this book, I try to convince the reader that even the origin of chloroplasts can be questioned. I provide various evidence and counterevidence to stimulate discussion on the origin of organelles.

The English version is a translation of the Japanese version by the author, but the two versions are not exactly identical. The original Chap. 1 is combined with Introduction (now Chap. 1 Introduction). Some episodes that are only understood by Japanese readers are omitted. The references in Japanese are kept with "J." Some paragraphs that are evident for English-speaking readers are simplified. Many figures are replaced by new, original ones to help comprehension. I hope the reader will enjoy this English version.

It may be relevant to note here on the distinction of the words, "theory," "hypothesis," and "discourse," which I used to qualify various arguments related to the endosymbiotic origin of organelles. "Discourse" is used to mean any argument, proposal, statement, or discussion, without judgment on whether they are correct or not. "Hypothesis" is a clearly identified statement proposed by representative researchers discussed in this book. It is sometimes wrong, and sometimes correct. "Theory" is a hypothesis recognized as correct and maintained for certain duration of time by the scientific community. Nevertheless, the distinction might not be clear in some cases, especially in which scientific discussions during the first half of the twentieth century are presented. "Discourses" will be used as the most general term throughout this book.

Contents

Chapter 1
Introduction

According to de Bary who first officially formulated symbiosis, "eine Betrachtung der Erscheinungen des Zusammenlebens ungleichnamiger Organismen, der Symbiose" ("symbiosis was a view on the coexistence phenomena of dissimilar organisms, the symbiosis," de Bary 1879, p. 5). He also noted: "Parasitismus, Mutualismus, Lichenismus v. s. v. sind eben jeweils bestimmte Specialfälle jener allgemeinen Associationseinrichtung, für welche der vorangestellte Ausdruck Symbiose als Collectivbezeichnung dienen mag." ("parasitism, mutualism, and lichenism are each a special case of general association states, for which previously defined expression symbiosis can be used as a collective notion," *ibid*, p. 21). He thought symbiosis notion was a new way of thinking, and added: "Was ich vorbrachte, enthält keine einzige neue Beobachtung, es sind lauter bekannte Dinge." ("what I told includes no new observations, but are all known facts," *ibid*, p. 30). Therefore, symbiosis is not a fact, but a notion that has its own history. The same is true for endosymbiosis, and endosymbiotic origin of organelles.

In this book, we will discuss theories on the endosymbiotic origin of organelles from the two points of view, namely, history of biology and current biological understandings. While keeping a prudent standpoint with respect to the theory of endosymbiotic origin of organelles, which still remain ambiguous in details, we will examine the history of endosymbiotic discourses as well as problems of the theory in the context of current, post-genomic biology.

There are two major topics in this book: First, who proposed the endosymbiotic theory of organelle origin, and what was the evidence for that? Second, in the light of current biological knowledge, how should we interpret the "endosymbiotic events"? To these ends, in Part I, we will trace the history of endosymbiotic discourses from their proposals to their acceptance, focusing mainly on the endosymbiotic origin of chloroplasts. We will examine the works of two major players, C. Mereschkowsky and L. Margulis, and discuss what their principal arguments were, and how these were accepted or refused. We will also see the status of the Mereschkowsky's theory in the biological community, and the opinions of various biologists around 1970 in which Margulis published her first comprehensive book

N. Sato, *Endosymbiotic Theories of Organelles Revisited*,
https://doi.org/10.1007/978-981-15-1161-5_1

on the endosymbiotic theory. All these reflections on the history of endosymbiotic discourses will lead us to think about who has the priority of the endosymbiotic theory. In Part II, I will first explain the basics of molecular phylogeny as a guide for understanding the evidence for the endosymbiotic theory. Then, we will see the importance of establishing the solid phylogenetic relationship beyond simple, apparent similarities of cyanobacteria and chloroplasts. Next, we will find some problems that are difficult to explain by a simple endosymbiosis, namely, the origin of enzymes for the chloroplast membrane biosynthesis and those involved in the chloroplast DNA replication. These problems arose in the post-genomic era, in which many genomic data became available for comparative genomics. With these discussions, I will present some problems in the present-day form of endosymbiotic theory on the origin of organelles, which we will have to solve. Finally, I will try to convince the readers to realize the importance of reconsidering the current, simplistic model of endosymbiotic origin of organelles.

To discuss endosymbiotic discourses on the origin of organelles, I will first explain symbiosis/endosymbiosis and organelles. In this chapter, I will explain these notions in plain words as a basis for further discussions. Endosymbiosis is a notion derived from symbiosis. Studies on symbiosis did not form a single discipline, having been performed in separate fields such as zoology, botany, and microbiology. Nevertheless, the work of a German biologist Paul Buchner (Buchner 1953) was a hallmark in the history of symbiosis studies. A Japanese book *Treatise on the Evolution of Cells* by Shichiro Sato (1988J) was also an interesting presentation of the author's original interpretations at the time.

1.1 Symbiosis

Currently, symbiosis is a term in ecology that designates the state of coexistence of two or more different organisms that have close interrelationships for survival. Ecological symbiosis is classified according to the relationship between the two (or more) organisms in the following ways: mutualism, in which both members gain benefits from the others, and commensalism, in which one of the members gains benefits one-sidedly. In many cases, symbiosis is understood in the sense of mutualism, but commensalism may be common. Two cases should be distinguished in commensalism, namely, whether the other member is harmed or not. In predation, a member is killed and eaten by the other. This is not normally included in symbiosis. Nevertheless, in an ecosystem in which only preys and predators exist, both populations oscillate periodically, and form an enduring system. This is well known theoretically and experimentally. We consider, however, symbiosis in common usage here. Representative examples of symbiosis are listed in Table 1.1.

Symbiosis of various organisms is not our main focus in this book, but it is important to understand that symbiosis is very common in the biosphere. Lichens served as a classical model of symbiosis consisting of multiple species of organisms. In animals including humans, microorganisms living in the digestive tract help digestion of foods

Table 1.1 Examples of representative symbiosis

Host	Symbiont	Classification	Explanation	Ref.
Lichens	Algae	Mutualism	Complex symbiotic systems consisting of fungi, algae, and cyanobacteria	1
Azolla (aquatic fern)	Cyanobacteria	Commensalism?	Cyanobacteria provide fixed nitrogen for *Azolla*	2
Symsagittifera (nematode)	Prasinophytes (green algae)	Commensalism?	Prasinophytes provide photosynthetic products for the nematode (photosymbiosis)	3
Herbivorous animals	Ruminant bacteria	Mutualism (digestive symbiosis)	Cellulose is degraded by the bacteria, which are used as proteins and fatty acids by the animal	4
Termites	Protists (*Oxymonas*, *Trichomonas*, etc.)	Mutualism	Cellulose is degraded by the protists to produce acetic acid and carbon dioxide. Acetic acid is oxidatively metabolized by the termites, whereas carbon dioxide and hydrogen are converted by the symbiotic bacteria within the protists to methane and acetic acid.	5
Mixotricha paradoxa (*Trichomonas* species)	Spirochetes	Commensalism?	Spirochetes are attached on the surface of *Trichomonas* cells within the termite digestive tracts, and help the cells to move	6
Land plants	Endophytes	Possibly mutualism	Exchanges of various substances are assumed	7
Land plants	Mycorrhizal fungi	Mutualism	The fungi provide phosphate, whereas the plants provide carbohydrates and fatty acids	8

Note that distinction between mutualism and commensalism is not rigorously defined (Keeling and McCutcheon 2017)

(1) Cain et al. (2002/2012), Section 3.5; (2) Watanabe (1992J); (3) Bailly et al. (2014); (4) various textbooks explaining ruminant bacteria and animal nutrition, such as Cain et al. (2002/2012), Chapter 27. See also Chapter 2 for enteric bacteria in various animals; (5) Kitade (2007J); (6) Cleveland and Grimstone (1964); Margulis (1970); (7) Reinhold-Hurek and Hurek (2011); (8) Luginbuehl et al. (2017)

Original table

in various ways. Many herbivores such as cows use ruminant stomachs. Rabbits have a long cecum. Humans, though omnivorous, have a long intestine. Many insects such as termites keep various microorganisms within their digestive tracts. We should recognize that the acquisition of digestive tracts was a revolutionary event in the evolutionary history of animals that established a symbiotic relationship between animals and microorganisms. In ecology, the food chain consists of producers, primary predators, secondary predators, detritivores, and decomposers. As the life of herbivores depends on the microorganisms in their digestive tracts, we have to consider a more complex, realistic situation, and understand that microorganisms are more than just decomposers.

Plants also harbor microorganisms called endophytes within their tissues. Endophytes somehow escape from the plant protection systems against infection, and perhaps exchange metabolites of signaling molecules with the host plants. Agronomical researches try to identify beneficial endophytes that promote plant growth, but there is much to study on what the endophytes do within the host plants.

1.2 Endosymbiosis

What is the difference between endosymbiosis and symbiosis? Endosymbiosis is a symbiosis within a eukaryotic cell, rather than within a body of multicellular organism. This distinction is important. Some old literature used the word endosymbiosis meaning the presence of exogenous cells within the spaces between the cells of multicellular organisms (this should be called intercellular symbiosis). For example, *Symsagittifera* (previously called *Convoluta*) harbor many prasinophytes within the intercellular spaces of its body (Table 1.1). However, we will use the term "endosymbiosis" in the sense in which a symbiont is present within a host cell. Representative examples of endosymbiosis are listed in Table 1.2.

1.2.1 Example 1: Corals and Zooxanthellae

One of the best-known example of endosymbiosis is zooxanthellae within corals (Fig. 1.1). Bleaching of corals is a serious, environmental problem, probably due to climate change. Corals might appear to be tree-like solid rocks, but are invertebrates harboring zooxanthellae cells (dinoflagellates) within their cells. Well-known endosymbiont species belong to the genus *Symbiodinium*, but only limited species are allowed for each coral species. Zooxanthellae cells also live as independent flagellates. Sea anemones and jellyfishes are also known as endosymbiotic hosts of zooxanthellae. In coral-zooxanthellae endosymbiosis, a mutualistic relationship is supposed, in which zooxanthellae cells provide photosynthetic products, whereas corals provide nitrogen source. Genomes of some *Symbiodinium* species were sequenced (about 600–900 MB), and mutual metabolite exchange was estimated (Lin et al. 2015; Aranda et al. 2016). However, quantitative measure of the contribution of the photosynthetic products of zooxanthellae to the host's nutrition is not clear.

1.2.2 Example 2: Root Nodules

Another well-known example of endosymbiosis is root nodules formed in the roots of legumes and other plants (Fig. 1.2). Nodules are cylindrical or spherical protrusion of about several millimeters formed in the root hairs as a result of infection by

Table 1.2 Examples of representative endosymbiosis

Host	Endosymbiont	Independent growth	Explanation	Ref.
Corals	Zooxanthellae (*Symbiodinium*, etc.)	Yes	Mutualism with exchange of carbon and nitrogen	1
Root nodules of leguminous plants	Nodulating bacteria	Yes	Mutualism with exchange of carbon and nitrogen	2
Paramecium bursaria	*Chlorella* (green algae)	Yes	Mutualism with exchange of carbon and nitrogen	3
Hatena	*Nephroselmis* (prasinophytes)	Probably no after ingestion	Prasinophyte chloroplast provides photosynthetic products for the host	4
Braarudosphaera bigelowii (Haptophyta)	Cyanobacteria (called UCYN-A in previous metagenomic analysis)	No	Cyanobacteria fix nitrogen and provide it for the host	5
Aphids	*Buchnera* (gamma proteobacteria)	No	Bacteriocytes in the body of aphid harbor *Buchnera* cells that provide amino acids to the host. The host provides some specific proteins for the bacteria	6
Cyanophora paradoxa (Glaucophyta)	Cyanelles (previously thought cyanobacteria)	No	Cyanelles are considered organelles but not endosymbionts	7
Paulinella chromatophora (Rhizopods)	Cyanelles or chromatophores (related to α-cyanobacteria)	No	Cyanelles are considered organelles but not endosymbionts	8
Rhopalodia, Rhizosolenia (diatoms)	*Cyanothece*-like cyanobacteria (spheroid bodies)	No	Cyanobacteria fix nitrogen and provide it for the host. Spheroid body should be considered an organelle	9
Sea slugs (*Elysia trisinuata*, etc.)	Siphonous green algae (*Codium fragile*, etc.)	Ingested chloroplasts do not proliferate	Undigested green algal chloroplasts called kleptoplasts are incorporated into the host cells and perform photosynthesis	10

(1) Thornhill et al. (2017); (2) Kawaguchi (2016J); (3) Hayakawa and Suzaki (2016J); (4) Inoue (2006/2007J); Yamaguchi et al. (2014); (5) Hagino et al. (2013); (6) Shigenobu and Wilson (2011); (7) Löffelhardt (2014), pp. 135–150; (8) Nowack et al. (2008); Löffelhardt (2014), pp. 151–166; (9) Nakayama et al. (2014); Löffelhardt (2014), pp. 167–182; (10) Yamamoto (2008J)
Original table

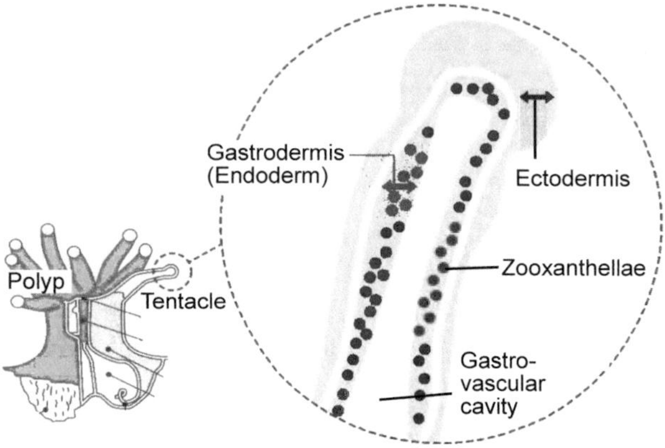

Fig. 1.1 A coral-zooxanthellae endosymbiosis. Original version of this figure was provided by the courtesy of Dr. Satoko Sekita (Kochi University)

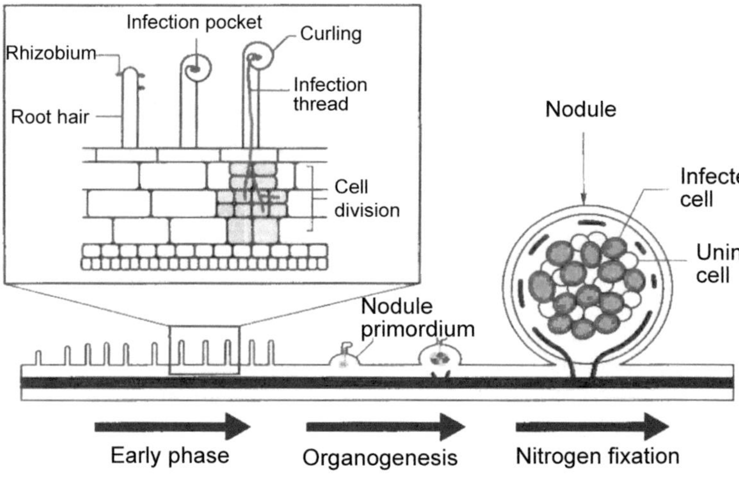

Fig. 1.2 Formation of a root nodule. Adapted from *Biology Illustrated* (2010J). Originally drawn by Masayoshi Kawaguchi (National Institute of Biology). Reproduced and translated with permission of the publisher and courtesy of Dr. Kawaguchi

nodulating bacteria, which grow as bacteroids within encapsulated inclusions in root cells. Nodules look pink or red due to leghemoglobins (products of plant tissue but not bacteria). The bacteria fix nitrogen by reducing nitrogen gas to ammonia, which is either provided for the plant tissues to synthesize amino acids, or converted within bacterial cells to amino acids which are subsequently transported to plant tissues. Plants fix carbon dioxide by photosynthesis in leaves, and the product (mainly sucrose) is transported through phloems to roots, and used as energy source for the growth of bacteria as well as for nitrogen fixation reaction. The enzyme involved in nitrogen fixation called nitrogenase is sensitive to oxygen. Leghemoglobin contributes to keep the interior of nodules free of oxygen. It is still unknown how nitrogen can enter the nodules while oxygen is excluded.

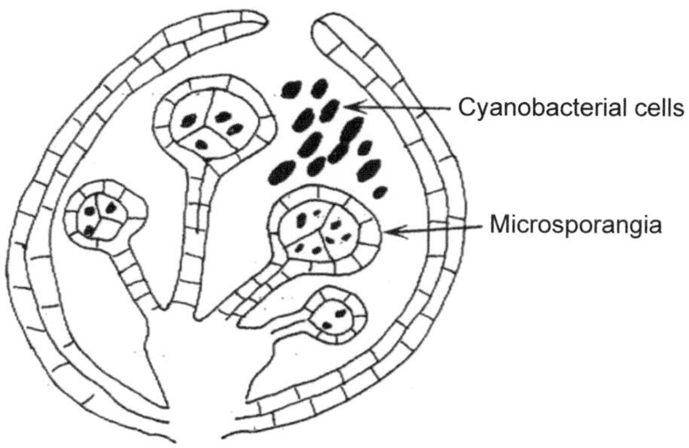

Fig. 1.3 Symbiosis of *Azolla* (fern) and *Anabaena azollae* (cyanobacterium). Adapted and reproduced from Watanabe (1992J) with permission of the publisher

As described, the relationship between nodulating bacteria and plants is a mutualism in which plants provide carbohydrates and bacteria provide ammonia. As a result, leguminous plants grow well on nutrient-poor soil with limited nitrogen nutrition. According to this property, leguminous plants are used as "green fertilizer," either in the rice field in early spring or on the ridgeway between rice fields.

A similar case (symbiosis, but not endosymbiosis) involving nitrogen fixation is the *Azolla-Anabaena* system. *Anabaena azollae* (cyanobacterium) cells grow within a hollow space in the rhizoids of an aquatic fern, *Azolla* (see Fig. 1.3). The fern provides carbohydrates, while cyanobacteria provide nitrogenous compounds. Note that this is not endosymbiosis, because the cyanobacteria are present outside the *Azolla* tissue.

1.2.3 Example 3: **Paramecium bursaria** *(green paramecia)*

This is again a classical example of endosymbiosis. A ciliate called *Paramecium bursaria* retains endosymbiotic *Chlorella variabilis* cells (called zoochlorellae) within a specialized vacuole called peri-algal vacuole (PV in Fig. 1.4). Protists like *Paramecium* incorporate bacteria and microalgae by phagocytosis, and digest them to use as nutrients. This particular *Paramecium* does not digest a particular species of *Chlorella* after phagocytosis, and keep it within the cell. The *Chlorella* cells proliferate by cell division within the *Paramecium* cell, which also grows and divide while harboring the *Chlorella* cells. *Chlorella* endosymbiont provides photosynthetic products (maltose), whereas *Paramecium* provides nitrogen source (ammonia). Note that green paramecium is different from *Euglena*.

As described above, endosymbionts are often photosynthetic organisms. This highlights importance of nutritional relationship in endosymbiosis.

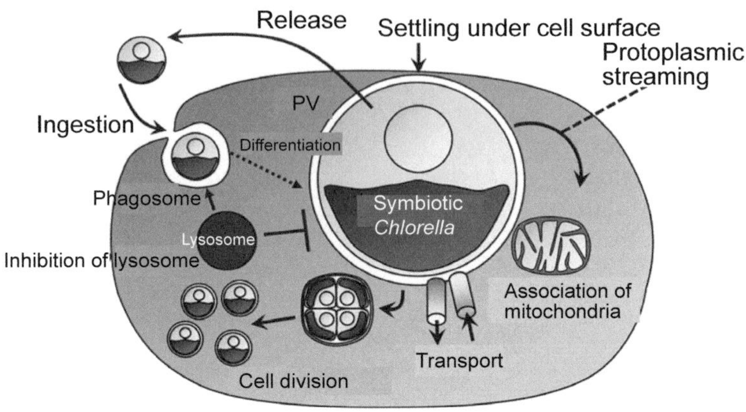

Fig. 1.4 Endosymbiosis of *Chlorella* cells in a green paramecium. PV, peri-algal vacuole. Adapted from Hayakawa and Suzaki (2016J) with permission of the publisher

1.3 Eukaryotic Cell and Organelles

Organelles are membrane-limited small particles present in eukaryotic cells (Fig. 1.5). Various organelles are known, but they are classified into two: those surrounded by double membranes, such as mitochondrion and chloroplast, and those surrounded by a single membrane, such as endoplasmic reticulum (ER), Golgi apparatus, vacuole (plants) or lysosome (animals), and microbody (also known as peroxisome and glyoxysome). Endosome is a vesicle formed by invagination of plasma membrane.

Why organelles exist in eukaryotic cells? Functional sharing may be a good idea. However, a single prokaryotic cell lives without similar functional sharing. We still do not know the exact reason. The presence of many mitochondria within a large eukaryotic cell is explained by the need for large area of membranes to enhance respiratory activity. Since respiratory chain is embedded in membranes, the hypothetical respiratory activity of plasma membrane (if respiratory chain existed in plasma membrane) can account for only a minor part of the energy demand of a large eukaryotic cell. A large number of mitochondria, each having intricately folded inner membranes, can provide enough energy for the eukaryotic cell.

Mitochondria and chloroplasts are distinct from other types of organelles in that they are surrounded by two membranes and contain their own DNA. They are equipped with complex genomic machinery that allows organellar DNA to function, namely, the systems of replication, transcription, and translation. The two organelles possess their own RNA polymerases and ribosomes, which are different from the counterparts in the nucleus and cytosol.

Now I will explain chloroplasts in detail, which will be the main topic in later chapters.

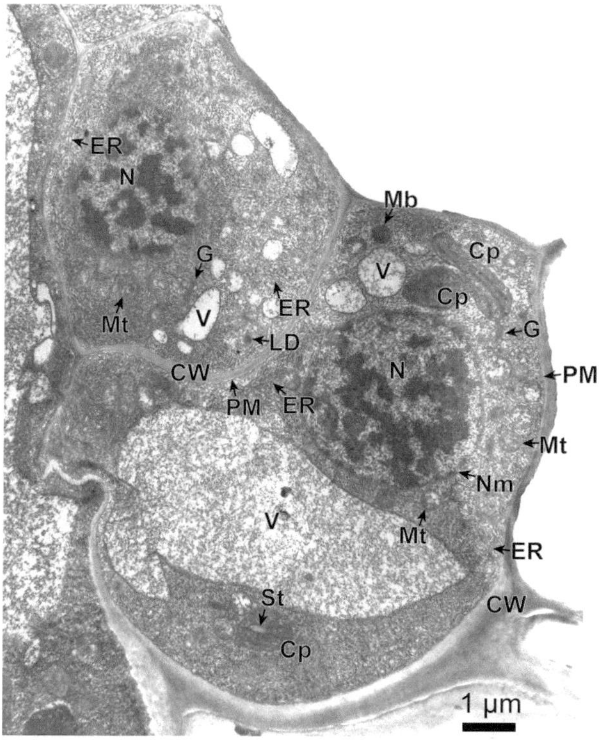

Fig. 1.5 Organelles in the plant cell. This transmission electron micrograph shows part of a leaf section of *Pisum sativum*. Two cells are visible in the center, while the cells on the right were lost during the preparation of section. Note that various organelles are tightly packed in the cytosol. *Cp* chloroplast, *CW* cell wall, *ER* endoplasmic reticulum, *G* Golgi apparatus, *LD* lipid droplet, *Mb* microbody, *Mt* mitochondrion, *N* nucleus, *Nm* nuclear membranes, *PM* plasma membrane, *St* starch granule, *V* vacuole. Original figure

1.4 Chloroplasts and Plastids

Chloroplasts are green granules involved in photosynthesis that are found in leaf cells of plants and algal cells. Photosynthesis is a process in which charge separation occurring in the chlorophylls (pigments) that absorb sunlight produces oxidants and reductants, the latter of which subsequently reduce carbon oxide to synthesize carbohydrates. This is not a place to discuss mechanism of photosynthesis, which is described in textbooks of plant physiology (Buchanan et al. 2002; *The Science of Photosynthesis* 2007J; Taiz and Zeiger 2010; Sato 2014J).

In most land plants, chloroplasts are present in green leaves and green stems. In orchids and some cultivars of radish, chloroplasts may also be found in some part of roots. Similar organelles devoid of chlorophylls are present in all nongreen tissues. They are classified into leucoplasts, chromoplasts, and etioplasts, among others. All these organelles are collectively called "plastids." In differentiating cells of the shoot apex, proplastids in undifferentiated cells are differentiated into chloroplasts. Leucoplasts are present in epidermal cells and root cells, whereas chromoplasts are in flower petals. It is important to note that any plant cell contain one type of plastids.

The interrelationship between these plastids was clearly described by a German botanist A. G. W. Schimper (1883, 1885). He showed that various plastids are related by inter-conversion or one-sided differentiation, and proposed: "plastids are produced only by plastid division." Later in the mid-twentieth century, development of plastids during greening was described in detail using electron microscopy (von Wettstein 1959). Figure 1.6 shows light-induced conversion of etioplasts to chloroplasts as an example. Other conversions, such as the conversion of chloroplasts to chromoplasts in petals and the conversion of proplastids to leucoplasts, are also known.

In 1980s, a microscopic technique of visualizing plastid DNA by fluorescent staining became popular in many laboratories. The plastid DNA within various algal cells and plant cells are now quite easily observed by fluorescence microscope (Fig. 1.7). DNA microfibrils can be occasionally observed by transmission electron microscope (TEM) (Fig. 1.8). In thin sections, we can see thin DNA fibrils in some blank area devoid of ribosomes or other particles. DNA may be abundant in the plastids (especially developing plastids), but is rarely observed as fibrils in thin sections. TEM was the only technique to observe organellar DNA in the 1960s, but currently fluorescence microscopy is much easier and sensitive for this purpose.

1.5 Endosymbiotic Theory on the Origin of Organelles

The origin of DNA-containing organelles with double limiting membranes, namely, mitochondria and chloroplasts, is explained by the endosymbiotic theory. Rigorously speaking, this should be "the theory of endosymbiotic origin of organelles." As "endosymbiosis" is a cytological event, this theory tries to explain the origin of organelles by hypothetical events of endosymbiosis that are supposed to take place long time ago. The notion "hypothetical events of endosymbiosis that are supposed to take place long time ago" is simply called "endosymbiosis." We will follow this convention in this book, but it should be noted that the same term is understood to mean a cytological event or an evolutionary process depending on contexts.

Although we will see details later, the theory of endosymbiotic origin of organelles (simply "endosymbiotic theory of organelles" below) assumes that both mitochondria and chloroplasts originated from free-living prokaryotes. Reasonable evidence for this theory is provided by the fact that both organelles contain their own DNA encoding some essential genes similar to bacterial genes as revealed by sequence comparison. Both chloroplasts and cyanobacteria perform oxygenic photosynthesis, which is another piece of good evidence for the endosymbiotic theory. Origin of mitochondria is supposed to be aerobic bacteria, whereas all kinds of plastids are believed to originate from cyanobacteria.

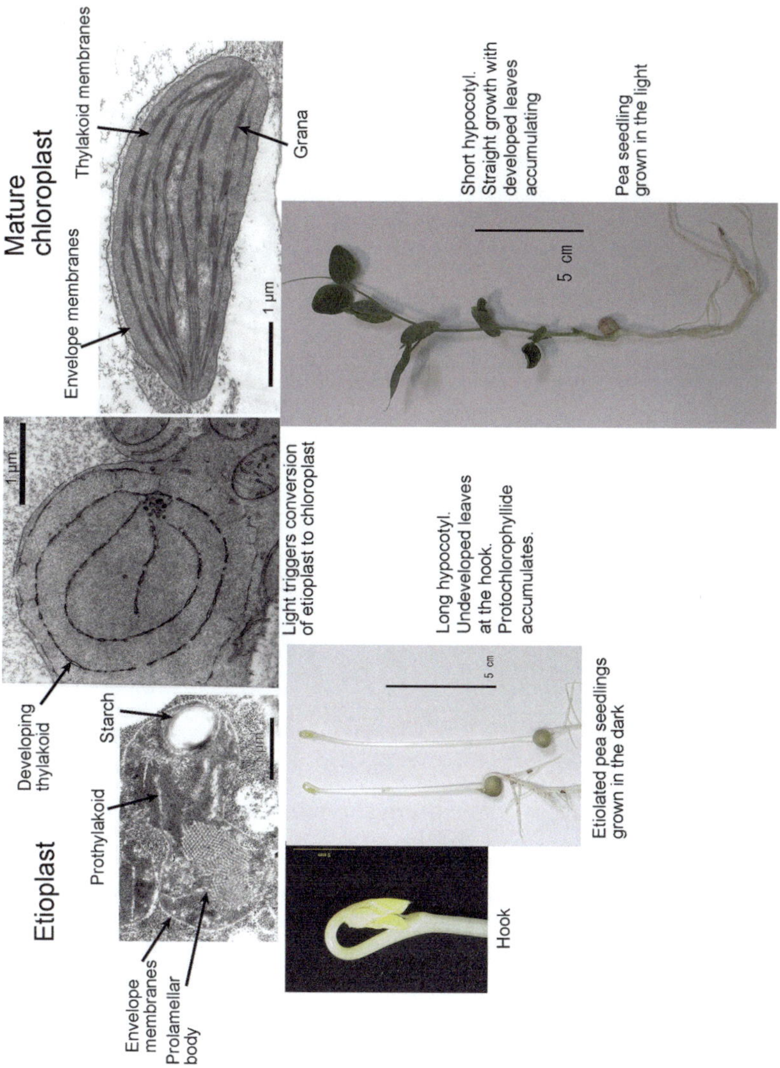

Fig. 1.6 Light triggers development of etioplast to chloroplast. Original figure

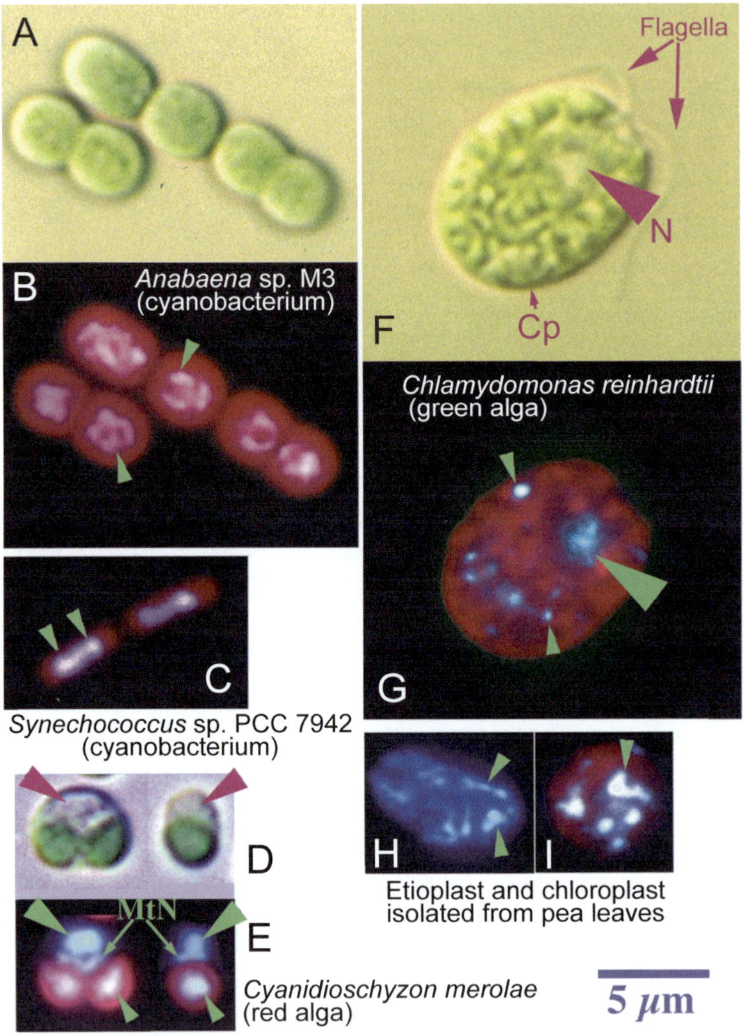

Fig. 1.7 Observation of nucleoids by fluorescence microscopy. (**a**, **b**) *Anabaena* sp. M3 (cyanobacterium); (**c**) *Synechococcus* sp. PCC 7942 (cyanobacterium); (**d**, **e**) *Cyanidioschyzon merolae* (red alga); (**f**, **g**) *Chlamydomonas reinhardtii* (green alga); (**h**) isolated pea etioplast; (**i**) isolated pea chloroplast. (**a**), (**d**), and (**f**) are differential interference, bright-field micrographs. Other panels are fluorescence micrographs showing DAPI fluorescence (blue) and chlorophyll fluorescence (red). Large arrowhead, cell nucleus; small arrowhead, nucleoid. *Cp* chloroplast, *N* nucleus, *MtN* mitochondrial nucleoid. Original in Japanese version except panels **c**, **h**, **i**, which were reproduced from Sato (2001) with permission of the publisher

Currently, mitochondrial origin is considered a rickettsia-like α-proteobacterium (or a closely related bacterium). The genus *Rickettsia* includes small parasitic bacteria that absolutely depend on their host organisms. It might be easy to understand a situation in which such parasitic bacteria gradually became organelles, but cur-

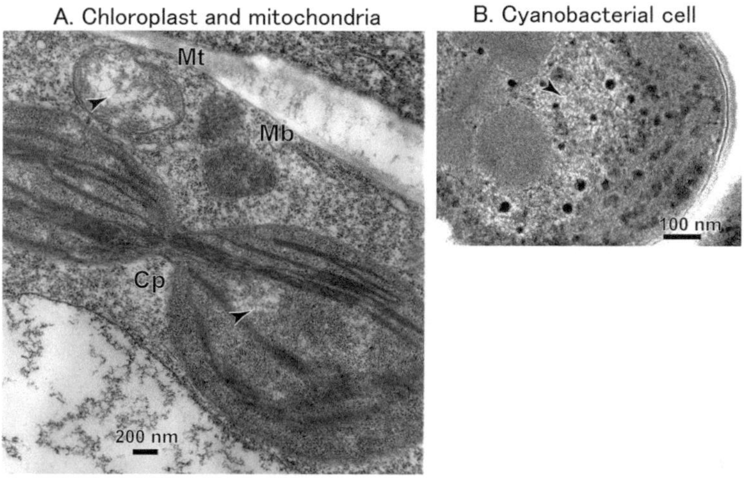

Fig. 1.8 Transmission electron micrographs showing DNA microfibrils. (**a**) Pea leaf cell; (**b**) *Synechocystis* sp. PCC 6803 cell. *Cp* chloroplast, *Mb* microbody, *Mt* mitochondrion. Arrowheads indicate DNA microfibrils. Original in Japanese version

rently known rickettsiae are all parasitic to human and animals. However, the ancestral rickettsia, endosymbiotic to a host cell that was an ancestor of all eukaryotic cells, might not be similar to the currently known ones. A deep branching cyanobacterium was supposed to be the origin of plastids, but the exact cyanobacterium that became endosymbiont is still unclear, because extant cyanobacteria and chloroplasts are indeed quite different.

As many readers know, the endosymbiotic theory is presented in high school textbooks (at least in Japan). Figure 1.9 is an example of illustration of the endosymbiotic theory. Similar diagrams are presented in many textbooks, such as *Campbell Biology* (Figs. 7–16 in Campbell et al. 2014/2016). It shows that a bacterial cell and a cyanobacterial cell were serially ingested into a host cell, which finally became a plant cell. Various specialist articles explain the details of the endosymbiotic processes (e.g., Fig. 2 in Bodył and Mackiewicz 2013; Figs. 1 and 2 in Archibald 2015; many figures in Maréchal 2018). In the original color illustrations of these articles (not shown here), the same color was used for the corresponding endosymbiont and organelle, which gives an impression that the endosymbionts became organelles without notable changes. The cyanobacterial cell and the chloroplast are drawn in a similar way, except for some details of the internal membrane systems. These visual images are certainly the origin of misunderstandings of many people interested in biology that chloroplasts are symbiotic cyanobacteria. Many researchers try to use, intentionally or unintentionally, reasoning that a particular process in chloroplasts must be similar to the corresponding process in cyanobacteria. It is true that homologs of some cyanobacterial genes exist in plants, and the gene products function in chloroplasts. The functional correspondence might be valid in some cases, but in other cases the functions of the homologs in cyanobacteria and chloroplasts are not identical.

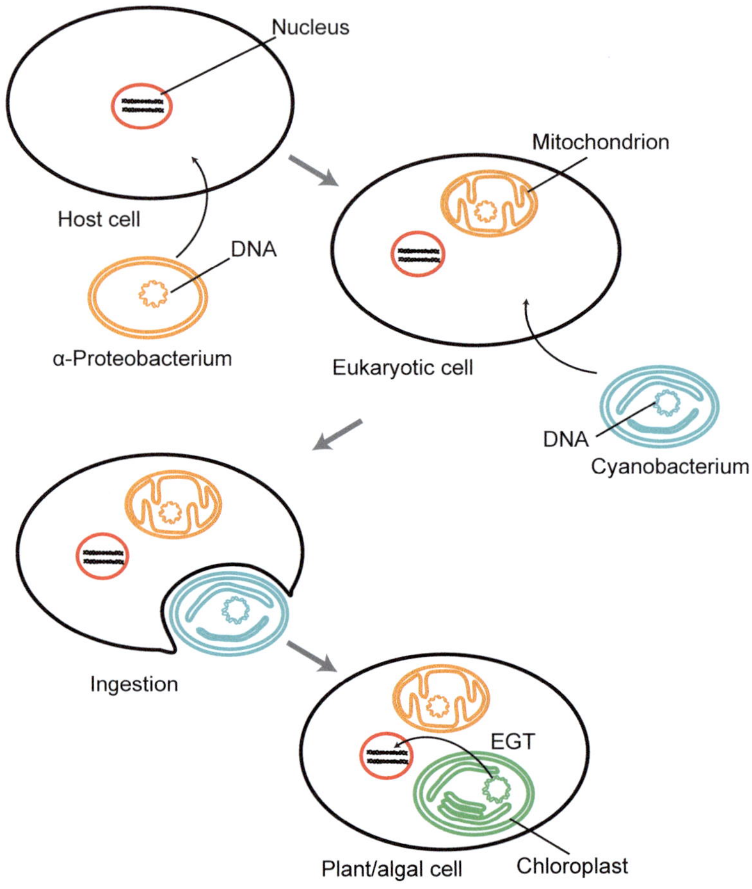

Fig. 1.9 A typical diagram showing the origin of mitochondria and chloroplasts. A similar diagram is found in many textbooks and review articles. Ingestion step for mitochondrion is not shown, but must be similar to that for chloroplast. Endosymbiotic gene transfer (EGT) is shown. Note that the cyanobacterial cell and the chloroplast are painted with an identical color in many textbooks and articles, which is avoided here. Note also that the nucleus is drawn smaller than the organelles to focus on the latters. Original figure

1.6 Fundamental Question: Are Membranes Inherited by Themselves?

However appealing the visual images of the endosymbiotic origin of organelles might be, we have a fundamental question on the images: namely, *membranes are not inherited by themselves*. The visual images consciously or unconsciously suppose that the envelope membranes and the thylakoid membranes are transmitted from the original endosymbiont to its descendants until present-day chloroplasts. The membrane hereditary hypotheses of Cavalier-Smith (2000), for example, explained the origin and heredity of various membranes of chloroplasts for easy understanding. Nevertheless, we should keep in mind that what is inherited is DNA (or genes) but not macroscopic or microscopic traits including cell structure. This is a fundamental of basic genetics. Then, how can we explain the apparent inheritance of chloroplast membranes? If the enzymes involved in the synthesis of the membranes (particularly, membrane lipids) are all encoded by the chloroplast genome, then we can safely say that chloroplast as a whole is inherited, and that the membranes are also inherited. As we will see later in Part II, this is exactly the situation in *Paulinella chromatophora*, which is hypothesized as a product of recent endosymbiosis. If, as some researchers believe, many chloroplast genes encoding the enzymes involved in the synthesis of membrane lipids are transferred to the cell nucleus by endosymbiotic gene transfer (EGT), we can still say that the chloroplast membranes are inherited as part of the chloroplast. But if the enzymes are simply encoded by the nuclear genome, we will have to examine the origin of the genes, namely, we will have to know whether they originate from the endosymbiont or are indigenous to the host cell. We will examine this crucial point in Part II. Without critical examination of the phylogenetic origin of the lipid synthesis enzymes, we cannot easily assume that the chloroplast membranes are inherited from the original endosymbiont. If the membranes are not inherited, how can we illustrate the image of endosymbiotic origin of chloroplasts? This is exactly the central question of the current book.

1.7 Necessity of Historical Analysis on the Establishment of the Endosymbiotic Theory

One of the major aims of the present work is to re-evaluate the endosymbiotic theory scientifically, as described in the previous section. At the same time, it is an important task to identify the biologist(s) who invented endosymbiotic theories in the history of biology.

Origin of Eukaryotic Cells (Margulis 1970) by Lynn Margulis (1938–2011) is a famous book that explained the endosymbiotic theory in detail from various different aspects. This epoch-making book was thought to give scientific evidence for the endosymbiotic theory, which remained hypothetical until then. Does anyone really read this book? As it is not translated into Japanese, I suppose that only a limited number of specialists read the book in Japan. *Symbiosis in Cell Evolution* (Margulis 1981/1993) that was translated into Japanese in 2002 might be more widely read among Japanese researchers and students. It is an enhanced version of the original 1970 book by adding new results of molecular phylogeny. The books of Margulis may be understood easily by just looking at figures, but the real contents are not easy to understand. Especially, the hypothesis on the origin of mitotic mechanism in eukaryotic cells, as well as the narratives on the origin of life in general that supported the main hypothesis, are really difficult to understand even for specialists in biology, because they are full of fantasy.

Upon looking at the reference list of the Margulis 1967 paper (Sagan 1967), we immediately understand that her narrative on the origin of life was a fruit of discussion with an astronomer Carl Sagan, who was her first husband. However, her narrative on the evolution of early life included some erroneous, nowadays unaccepted knowledge, such as the idea that aerobic bacteria appeared after cyanobacteria as an adaptation. The explanation of the origin of mitosis in eukaryotes is hard to understand for current biologists, and the same must be true for specialists of the time. In the light of current biological knowledge, her explanation was entirely false. However, she did not give up this peculiar hypothesis in her 1998 book *Symbiotic Planet* or her later publications. Many readers may perhaps skip this problem as a black box, but the origin of mitosis was the core of her idea, as I will explain in detail later. Without deciphering this core, we will not be able to evaluate the contribution of Lynn Margulis to the endosymbiotic theory of organelles.

Then, who else invented the endosymbiotic theory? The articles reporting the discovery of DNA in chloroplasts and mitochondria in 1960s already cited original papers on the endosymbiotic origin of chloroplasts, written by two Russian botanists. They were Constantin Mereschkowsky (1855–1921) and Andrei Famintzin (1835–1918). Note that several different spellings in Western languages were used for these authors in different literatures. Their papers on the endosymbiotic origin of chloroplasts were published not only in Russian journals but also in a German journal, *Central Blatt*, one of the major biological journals of the time, and were widely read. However, their hypothesis remained just one of the possibilities, because very limited knowledge on the basic metabolism and biochemical substances was available in the first decades of the twentieth century. I am afraid that not many researchers (except German and Russian scientists) have read their papers, because non-English literature is no longer considered in current biology and very limited number of researchers can read old articles written in German and Russian. Although she was not able to read papers in these languages, Margulis cited sometimes papers of Mereschkowsky and Famintzin in incomplete or wrong forms. Judging from these circumstances, I describe Mereschkowsky as another major player in this book. His paper on the proposal of endosymbiotic origin of plastids (Mereschkowsky 1905a) was translated into English (Martin and Kowallik 1999) and Japanese (Sato 2016J). If we compare Mereschkowsky's idea with the narratives of Margulis, we will be able to understand how the endosymbiotic theory was developed and established, which is a prerequisite for discussing the logical construction of the theory and its significance in contemporary biology.

In Part I, I will present the hypotheses of Mereschkowsky and Margulis, as well as various related discourses. Re-examination of the endosymbiotic theory from the viewpoint of contemporary biology will be the main topic in Part II.

1.8 From Similarity to Continuity, Then to Discontinuity

I should like the readers to notice that there are two meanings of endosymbiosis in the endosymbiotic theory of chloroplast origin: first, endosymbiosis is understood as a cytological event, in which a cyanobacterial cell is incorporated into a eukaryotic host cell. Second, it is an evolutionary hypothesis, in which chloroplasts originate from a cyanobacterial ancestor. These two meanings are usually mixed in an ambiguous image. We should recognize the principle of "continuity of chloroplasts," which states that chloroplasts are derived from chloroplasts. Curiously, in the discourses of the current biology, we suppose, or we are forced to suppose, the continuity of cyanobacteria and chloroplasts in the evolutionary history. To fit this idea, we are tempted to assume perfect similarity of cyanobacteria and chloroplasts. Common properties of cyanobacteria and chloroplasts may be regarded as reliable supports for the continuity of the two entities. This type of argument was exploited in historical endosymbiotic hypotheses, that I will explain in Part I.

In contrast, recent phylogenetic studies as explained in Part II began to indicate that the cyanobacteria-chloroplast continuity must be considered as significantly limited than it was thought before. This is not a problem of similarity and difference, but we should recognize the situation in which similar traits do not always originate from an identical resource. In other words, we are discovering discontinuity. Different interpretations of the discontinuity will lead to different evaluation of endosymbiotic hypotheses. One might think that there was a single endosymbiotic event, but that there were additional, many lateral gene transfers. One might also think that there were as many gene transfers as exogenous genes, including the genes that have been considered to originate from the endosymbiont. A contrasting idea may be that there were as many endosymbiotic events as exogenous genes. In one idea, exogenous genes could have entered the host cell before the primary endosymbiosis. In another, they could have replaced the genes introduced by the primary endosymbiosis one by one. We can imagine various possible ideas like these, which are currently discussed extensively. The origins and histories of mitochondria and chloroplasts might be quite different. We still do not know how the eukaryotic cell containing these organelles was born. Various different versions of endosymbiotic discourses, closely linked to the hypotheses on the origin of eukaryotic cell, remain probable with a combination of two parameters: first, continuous lineages linking organelles and their bacterial origins; second, interpretations on the discontinuities that are found by experiments and phylogenetic analyses.

Under these situations, it is timely to reflect ourselves on the past history of endosymbiotic discourses to recognize why we had to assume and accept the endosymbiotic theory as an explanation of the origins of organelles. It is also important to ask ourselves if we have accumulated puzzling interpretations to ignore the discontinuities without actually re-examining the process in which we accepted a very naive version of the endosymbiotic hypothesis. This prompted me to write this book, combining history and biology. Now, we shall start with history.

1.9 A Short History of Endosymbiotic Theories

Before describing details, we will briefly overview the whole history of the endosymbiotic discourses, mainly focusing on the endosymbiotic origin of plastids, namely, the initial proposal of the hypothesis, variations of the original hypothesis and counterarguments, and unanimous acceptance of the theory. I intentionally set aside discourses on the endosymbiotic origin of mitochondria in the Japanese version, because I thought that only limited sources of evidence have been provided for the mitochondrial origin. Indeed, there are various forms of mitochondria, some having DNA and others without DNA, some are aerobic and others are anaerobic. In this English version, I have added brief discussions on this problem in Sect. 6.5. The readers are also referred to interesting reviews and books on this aspect (Martin and Müller 2007; Martin et al. 2015; Archibald 2014, 2015).

Although the following is a short chronology on the endosymbiotic discourses of plastid origin, a list of references in the chronological order at the end of this book will be also useful as a guide for the historical evolution of endosymbiotic discourses.

1883	Schimper suggested endosymbiotic origin of plastids as a possibility in the context of continuity of chloroplasts and plastids
1905	Mereschkowsky proposed that plastids originated from ancestral cyanobacteria by endosymbiosis (multiple parallel endosymbioses)
1907	Famintzin proposed an endosymbiotic theory of plastids
1910, 1920	Mereschkowsky published two long articles on his endosymbiotic theory
[Meanwhile, endosymbiotic hypothesis remained discourses of a limited number of authors]	
1952	Lederberg formulated genetic symbiosis, including endosymbiosis and virus infection
1962	Ris and Plaut discovered chloroplast DNA in *Chlamydomonas reinhardtii*, and suggested a possibility of endosymbiotic origin of chloroplast, citing the papers of Mereschkowsky and Famintzin.
1963	Nass et al. discovered mitochondrial DNA
[Serious discussion began in academic societies on the endosymbiotic origin of organelles]	
1966–1967	Short papers proposing endosymbiotic origin of plastids were published
1967	Lynn Margulis (Sagan at the time) published an article on the endosymbiotic origin of mitosis in eukaryotic cells. She tried to explain the origin of eukaryotic cells by endosymbiotic origin of mitosis, mitochondria and chloroplasts (multiple parallel endosymbioses)
1969	In the Symposium of the Society for General Microbiology, Stanier reviewed the current status of knowledge on the endosymbiotic origin of organelles. The proceedings were published next year
1970	Margulis published *Origin of Eukaryotic Cells*
1967–1975	Various articles supporting autogenous (continuous) and endosymbiotic hypotheses were published
1974	Stanier supported cyanobacterial origin of red algal plastids

(continued)

1975	Discovery of a prokaryotic green alga *Prochloron* supported the hypothesis of multiple parallel endosymbioses (Prochlorophytes were identified as polyphyletic groups within cyanobacteria in the 1990s)
1975–1978	Molecular phylogeny supported the endosymbiotic origin of plastids. Mitochondrial origin remained uncertain
1981	Whatley proposed the notion of secondary endosymbiosis to explain the origin of Cryptophytes, Alveolates, etc.
1982	Gray published a review affirming the endosymbiotic origin of plastids. He was still cautious about the endosymbiotic origin of mitochondria, which will be accepted in his 1992 review
2000 onwards	Detailed analysis on the endosymbiotic origin of organelles began based on large-scale genomic data (single origin of organelles, identification of endosymbiont, etc.)

The chronology listed above shows that the presence of organellar DNA was the key evidence for the endosymbiotic theory. Remaining problems reside in other characteristics of chloroplasts, such as the mechanism of photosynthesis, genetic system, and membrane biosynthesis. How did these affect the endosymbiotic theory? Did they support it? Or, they seemed to support it but finally they might not. In parallel, it is also important to identify who contributed to the endosymbiotic hypotheses, and why some researchers changed their opinion with time in the history.

I already presented representative publications of Lynn Margulis, as well as some important monographs on the endosymbiotic theories. For the entire history of biology, a French book by Morange (2016) that I recently translated into Japanese is insightful.

Part I
History of Endosymbiotic Discourses

In this part, we will review who proposed which forms of endosymbiotic discourses, and we will discuss development in biology behind various forms of endosymbiotic discourses. Old history of these discourses, full of unfamiliar names and notions, might be mostly unknown to young scientists and students. However, many of the scientists that we will talk about were certainly influential at their time. It is essential to recognize that many researchers before Lynn Margulis were seriously thinking about the endosymbiotic hypothesis with some convincing evidence.

Chapter 2
Mereschkowsky, Founder of Endosymbiotic Hypothesis

In this chapter, we will overview the life and works of a Russian biologist Constantin Mereschkowsky. As described already, he was the first scientist that proposed the endosymbiotic hypothesis in a clearly defined form. He is sometimes described as an eccentric man, but his talent invented an original hypothesis, which is still echoing in the current theory of endosymbiosis in an interesting manner.

First, I would like to explain the background knowledge that link Margulis and Mereschkowsky. Margulis (1970) described that the endosymbiotic origin of plastids (chloroplasts, leucoplasts, and chromoplasts) was "easiest to defend" (p. 278). The most important evidence for this was the discovery of chloroplast DNA by her previous supervisors, Hans Ris and Walter Plaut (Ris and Plaut 1962). This paper cited Mereschkowsky (1905a) and Famintzin (1907) as the first articles that proposed the endosymbiotic origin of chloroplasts. Margulis used Wilson's textbook of cytology (Wilson 1925) in the lecture of Ris, and was familiar to the endosymbiotic theory, although Wilson presented the theory in a very negative, mockery attitude: "Mereschkowsky, in an entertaining fantasy, has developed the hypothesis" (p. 738). Margulis (Sagan 1967) indirectly cited Mereschkowsky in her 1967 paper through Wilson's textbook, certainly because she did not read the original article. To evaluate the originality of the works of Lynn Margulis, it is essential to clarify what had been known and said before her. We have to examine what Mereschkowsky, known as the first Russian advocate of the endosymbiotic hypothesis, wrote exactly on the hypothesis.

For this reason, I have translated Mereschkowsky's 1905 paper (in German) into Japanese, and published it with short explanations in News Letter of the Japanese Society of Photosynthesis Research (Sato 2016J). The same paper was translated into English by Martin and Kowallik (1999), which, I regret, may not be easy to understand for young students because many phrases included direct translation of old words and notions. Here I will explain the life and works of Mereschkowsky in plain words, to convince the readers of the historical importance of his works.

N. Sato, *Endosymbiotic Theories of Organelles Revisited*,
https://doi.org/10.1007/978-981-15-1161-5_2

2.1 Mereschkowsky's Life

The life of Constantin Mereschkowsky (1855–1921) was presented in detail in Sapp et al. (2002), which also included a complete list of publications. The Russian book of Khakhina (1979/1992) that was later translated into English under the leadership of Lynn Margulis also described the life and works of Mereschkowsky.

C. Mereschkowsky was born in St. Petersburg in 1855. He was accepted in St. Petersburg University in 1875 and began studying zoology. After graduation in 1880, he worked as a research assistant for some years, and then visited institutes in Paris, Roscoff, and Berlin. In Berlin, he became acquainted with famous biologist Rudolf Virchow (Sapp et al. 2002). Studies on animal pigments like butterfly scales led him to get a post of "privatdocent" (lecturer) in 1883. However, he moved to Crimea in 1886, and then visited various institutes. From 1898 until 1902, he stayed in California in USA to study classification of diatoms and wrote some papers on the topic. He returned to Russia to be a curator of zoological museum in Kazan Imperial University. He was given master's degree in botany for the studies on diatoms in 1903. He then became a privatdocent (lecturer) of botany in 1904. He was interested in the problems of symbiosis, and studied lichens. In 1906, he received doctorate in botany, and then became a professor in botany in 1908. Nevertheless, he was obliged to leave the post due to his pedophilia scandals as well as political status. He moved to Nice in France and then to Geneva in Switzerland. After publication of his large paper (Mérejkovsky 1920), he considered that he did everything he should do, and he died in 1921 with his complicated large machine for suicide (Sapp et al. 2002).

It is curious to everyone that Mereschkowsky was just a privatdocent at the age of 50, when he published a paper on the endosymbiotic hypothesis of plastid origin. Perhaps for this reason, there was an image that an ill recognized scientist in a local university far from the center of European science courageously submitted a manuscript of his unprecedented theory to the major German journal of biology (see, e.g., Martin and Kowallik 1999, p. 287). However, Kazan University was one of the most important imperial universities in Russia. Mereschkowsky visited many times various institutes in Europe and the USA, and published various papers on his research, although his frequent stays in foreign countries was due to unfinished troubles that he made.

After his 1905 paper, Mereschkowsky became a professor, and published a large paper in German (Mereschkowsky 1910). The paper he wrote in French before his death was really a large treatise on his endosymbiotic theory (Mérejkovsky 1920). It is natural that he also wrote papers in Russian (a complete list is provided in Khakhina 1979/1992), but, as far as I judge from the titles, they were papers on the same topics as the corresponding papers in German or French. Under these circumstances, he was not a neglected scientist, in contrast with prevailing prejudice or imagination on his status in science. Precisely, his hypothesis was well known in scientific community, but was subject to caution and skepticism. Sapp et al. (2002, p. 436) denied, however, the originality of Mereschkowsky, who "selected among them (all elements of his theories scattered in the scientific literature) to see how the pieces could be fit together", in other words, simply assembled various data of the time into a postulate of endosymbiotic origin of plastids and origins of life. I do not share this view. Sapp, a Canadian historian, perhaps evaluated too much the importance of endosymbiotic theory in the 1980s, supported by the molecular phylogenetics mainly advanced in Halifax, Canada, by his compatriots.

In a different perspective, symbiotic notion characterized Russia of the time: an anarchist Peter Kropotkin worked for the symbiosis in human society, while Sergei Winogradsky (or Vinogradsky, 1856–1953) studied ecology of microbial world (Morange 2016). It is quite probable that the Russian background enabled Mereschkowsky to conceive and develop endosymbiotic notion of plastid origin.

2.2 Mereschkowsky's Hypothesis on the Endosymbiotic Origin of Chloroplasts

Mereschkowsky described in his 1905 paper entitled " Über Nature und Ursprung der Chromatophoren im Pflanzenreiche" (On the nature and origin of plastids in plant kingdom) that plant and algal cells were generated by an endosymbiosis of a blue-green alga (now called cyanobacterium) into an animal cell and that the continuity of plastids (every plastid is generated from a plastid) was established thereafter (Mereschkowsky 1905a, b). Since the name "cyanobacterium/cyanobacteria" (singular/plural) was established recently (in 1980s), we use "blue-green alga/algae" or "Cyanophyta, Cyanophyceae" in historical descriptions in this book. The continuity of plastids had been established by Schimper (1885), and gave inspiration to Mereschkowsky on the endosymbiotic origin of plastids.

Here I should add a short comment on the term "plastid," which is an established notion in modern biology. At the time of Schimper and Mereschkowsky, plastid was also called chromatophore. Chromatophore is now used for internal membrane inclusions of anoxygenic photosynthetic bacteria. It is also used to denote chloroplast-like organelle within a photosynthetic rhizopod *Paulinella chromatophora*. We try to avoid use of chromatophore in the sense of plastid in this book. If necessary, a note will be added. Original words are shown in parenthesis as necessary.

In this section, I will present the outline of Mereschkowsky (1905a, b) with my short annotations. The logic of Mereschkowsky was the following (note that each header is the same as the one presented in his papers):

2.2.1 Continuity of Plastids ("Chromatophoren" in the Original Text)

Mereschkowsky reasoned like this: Schimper established this notion, in which plastids are generated only by division of a plastid. In other words, no plastids arose spontaneously de novo. If we analyze in this way, plastids existing in the past were the ancestors of all existing plastids. If we consider the origin of life, plastid must have been created at a time in the past. If it was produced spontaneously from the constituents of the cell within the cell in its creation, we can imagine that similar events must occur even now. This is not true. Then, plastids must have emerged at a fixed time point. A rational explanation for this can be given by a hypothesis that a free-living photosynthetic bacterium entered the cell of an ancestor of plants to become a chloroplast.

Mereschkowsky thus derived the endosymbiotic hypothesis of plastid origin by pure logic.

2.2.2 High-Level Independence of Plastids from the Cell Nucleus

At the time of Mereschkowsky, it was technically possible to remove a part of protoplasm of plant cell by protoplasmic separation, and allowed the plastids therein to survive. In addition, once isolated, plastids were known to survive for some time. These observations were taken as evidence for the partial independence of plastids from the cell nucleus. In current biology, isolated chloroplasts are used for photosynthetic researches and gene expression experiments. It is also true that many chloroplast proteins are encoded by the nuclear genome, and that the chloroplast function is maintained by the transport into the chloroplasts of precursor proteins synthesized in the cytosol. In other words, chloroplasts are not completely independent from the cell nucleus, but behave like independent entities within a limited time period.

Mereschkowsky did understand this limitation. He reasoned that chloroplasts are no longer completely independent from the cell nucleus because long time passed after the establishment of chloroplasts within the plant cell.

2.2.3 Perfect Resemblance ("Analogie") of Plastids with Zoochlorellae

Mereschkowsky stated that *Amoeba viridis* harbors parasitic chlorella cells, which are considered highly independent, because they grow and divide like plant plastids. Famintzin criticized this, based on the essential distinction between chlorella and plastids, because chlorella is a eukaryote, clearly different from prokaryote-like plastids, although the distinction between prokaryote and eukaryote was introduced later. However, we know that Mereschkowsky clearly recognized this distinction. He used this comparison because he tried to present examples in which photosynthetic microorganisms are symbiotic. Another example he cited was zooxanthelles. It is important to know, however, that, once isolated, plastids do not survive, whereas chlorella cells continue to live autonomously. As explained earlier, Mereschkowsky explained this difference in terms of difference in time after the beginning of symbiosis.

2.2.4 There Are Organisms That We Can Consider Free-Living Plastids

Mereschkowsky asserted that plastids are similar to unicellular, free-living blue-green algae, especially similar in cell structure. We will summarize similarities of cyanobacteria and plastids in the next section. In the time of Mereschkowsky as well as the time of Margulis, the notion symbiosis curiously included two modes of life, namely, free-living state and symbiotic state. For this reason, the statement that plastids are symbionts presupposed the presence of free-living organisms that are similar to plastids.

2.2.5 Blue-Green Algae ("Cyanophyceen") Virtually Live as Symbionts Within the Cell Protoplasm

This is a complementary situation to the previous point. Mereschkowsky listed three examples in which free-living blue-green algae are trapped within a cell, and maintained for a long time. We will come back to them in Sect. 2.4 of this chapter.

2.3 Comparison Between Blue-Green Algae and Chloroplasts

In this section, we will examine the comparison of blue-green algae and plastids according to Mereschkowsky (Table 2.3). A current view on this comparison will be presented in Part II.

Although distinction between eukaryotes and prokaryotes was not established at the time, Mereschkowsky did understand "authentic nucleus," and considered both blue-green algae and plastids as prokaryotic (Item 3). This is clearly different from the green algal endosymbiont in green paramecia. We should recognize that

Table 2.3 Comparison of blue-green algae and plastids according to Mereschkowsky (1905a, b)

Cyanophyceae (*Aphanocapsa*, *Microcystis*, etc.)	Plastids
1. Small, homogeneously green, spherical, or ovoid particles, having a very simple structure	1. Small, homogeneously green (originally, blue-green like *Cyanomonas*), spherical, or ovoid particles, having a very simple structure
2. Green pigment is distributed uniformly in the protoplasm, or as very small droplet-like forms	2. Green pigment is distributed uniformly in the stroma, or as very small droplet-like forms
(version 1905a) 3. No authentic nucleus is present. Instead, only a certain structure (Nucleinkörper) is present that can be considered as precursor to the nucleus	3. No authentic nucleus is present. Instead, only certain structures (pyrenoids) are present that can be considered as deformed, primitive nucleus
3. No authentic nucleus is present. Instead, only a certain central mass (Zentralmasse), probably invisible small chromatids, is present	3. No authentic nucleus is present. Instead, only eventually a certain colorless central mass (pyrenoid) is present that can be considered as a homolog of cyanophycean central mass
(version 1905a) 4. Nutrition: Assimilation of carbon dioxide under light	4. Nutrition: Assimilation of carbon dioxide under light
4. Nutrition: Assimilation of carbon dioxide under light. Synthesis of protein	4. Nutrition: Assimilation of carbon dioxide under light. Synthesis of protein
5. Proliferation: By division	5. Proliferation: By division

The corrected version 1905b is shown along with the version 1905a

Mereschkowsky used eukaryotic algal examples to explain endosymbiosis of blue-green algae. In retrospect, however, the nucleus-like structure in item 3 was pyrenoid in chloroplasts and carboxysome in blue-green algae. It is improbable that DNA-containing nucleoids were visible then. What is important here is that he recognized the lack of authentic nucleus in plastids and blue-green algae.

Items 1 and 2 described similarity of microscopic characters. Although this seem to be quite superficial, Mereschkowsky pointed out homogeneous distribution of green materials, therefore, of thylakoid membranes (not known then), in both blue-green algal cells and chloroplasts. We still do not understand if the small droplets represented grana or other granules.

Item 4 was the most important in the table. Photosynthetic activity was listed as a similarity of blue-green algae and plastids. This is the most important item that could have been listed first, but Mereschkowsky preferred to list morphological characters first and then physiological characters, perhaps according to the tradition of the time. After revision in Mereschkowsky (1905b), the author added protein synthesis in item 4. Obviously, it was not technically possible to detect protein synthesis within the plastids and blue-green algae. This could have been a result of pure inference. This corresponds to advances in biochemistry, just establishing as a discipline independent of organic chemistry, which emphasized the importance of protein (Morange 2016).

Item 5 is also an important statement of similarity, in which both blue-green algae and plastids proliferate by binary fission, which is different from the division of eukaryotic cells. In plant cell division, a cell plate is formed in the future division plane within the cell, and enlarged to form a new cell wall. In this regard, plastid division was considered similar to the division of blue-green algae. Many prokaryotic factors in chloroplast division were indeed discovered from around 2000, and the real similarity of the two division systems was discovered (Miyagishima 2005). The knowledge on the similarity of blue-green algae and plastids in 1970s was, however, not very different from the one in the beginning of the twentieth century. The same is true for other items in Table 2.3. The only knowledge of 1970s that was not present in the early 1900s was the presence of chloroplast DNA, which was discovered in 1962. This will be discussed later.

2.4 Implications of Mereschkowsky's Examples of Living Blue-Green Algae Within Algal Cells

Mereschkowsky showed the following examples of living endosymbionts within the cells. As the description of Mereschkowsky was incorrect, I present additional information from the AlgaeBase (http://www.algaebase.org) as follows:

- *Paulinella chromatophora* (rhizopod, Lauterborn 1895).
- *Cryptomonas americana* (cryptophyte, Davis 1904. Mereschkowsky described "Davis 1894" without details of references).
- *Rhizosolenia styliformis* (diatom, it was originally discovered by Brightwell 1858, but Mereschkowsky cited "Schmidt 1901." He spelled "Rhixosolenia.")

These were the newest discoveries at the time relying on the newly identified organisms. Intriguingly, these were recently identified as interesting examples related to endosymbiotic origin of plastids, and thus played important roles in the history of endosymbiotic hypothesis. Although Mereschkowsky regarded these as the real endosymbiotic systems including living symbionts, they are different from original blue-green algae, and are not real symbionts living independently from the hosts. We will examine them individually.

Paulinella chromatophora, together with *Cyanophora paradoxa*, has long been regarded as a typical example of endosymbiosis involving true blue-green algal endosymbionts, which have been called cyanelles (two writings are used: cyanelle–cyanelles, and cyanella–cyanellae, for singular–plural, respectively). We find traditional explanations of cyanelles in the famous textbook of Roger Stanier and others:

> Blue-green bacterial symbionts are called *cyanellae*. They are found in a few genera of freshwater protozoa (e.g., the flagellates, *Cyanophora* and *Peliaina*, and the ameboid rhizopod, *Paulinella*).

Peliaina contains from one to six cyanellae. The symbiosis is maintained by balanced cell division, but this mechanism sometimes fails and a cyanella-free protozoan cell is formed. In *Cyanophora* and *Paulinella*, however, the symbiotic association is perfectly regulated: the protozoan host usually contains two cyanellae, each daughter protozoon receiving one cyanella at cell division. The symbiont then divides, restoring the number of two per host (Stanier et al. 1957/1976, p. 758).

Peliaina cyanea is a glaucophyte like *Cyanophora paradoxa*, and its cyanelle was classified as a unicellular blue-green alga belonging to the genus *Synechococcus*. The description of *Peliaina* was based on Pascher (1929b), and Myer-Abich (1950) also presented a similar description. *Cyanophora paradoxa* was discovered by Korschikov (1924), 3 years after the death of Mereschkowsky. The cyanelles in *Paulinella chromatophora* were also called chromatophores, which were isolated by Lauterborn (1895) and Pascher (1929a), but do not survive for a long time after isolation. According to the contemporary phylogenetic analysis, all plastids are monophyletic, having a deep root in cyanobacteria, whereas the chromatophore originated from an α-cyanobacterium (*Prochlorococcus* and marine *Synechococcus* group), and therefore regarded as a result of another endosymbiotic event (Nowack et al. 2008). As explained above, the term chromatophore was used to designate plastid in early twentieth century. Further details of chromatophores of a rhizopod *Paulinella chromatophora* will be presented in Sect. 6.4.

Cryptophytes including *Cyanomonas* were the first secondary endosymbionts proposed in the 1980s. They have not only a plastid but also a reduced nucleus (nucleomorph) of the presumptive endosymbiotic red alga (Douglas et al. 2001).

Some diatoms such as *Rhizosolenia* contain, within the cytoplasm, a colorless granule called spheroid body, which was similar to blue-green algae in the eyes of Mereschkowsky. This was a marvelous foresight of Mereschkowsky, because we can only understand this similarity by electron microscopy. This was certainly a fruit of his studies on diatoms in California. According to recent findings with genomic analysis, spheroid body is involved in nitrogen fixation rather than photosynthesis (Nakayama et al. 2014). In this respect, this represents the third event of cyanobacterial endosymbiosis. In retrospect, many suggestive ideas were found in the discourses of Mereschkowsky.

After the 1905 papers, Mereschkowsky published two long articles explaining his hypothesis in European biological journals (Mereschkowsky 1910; Mérejkovsky 1920). The different spellings reflected different ways of transcription of Russian name in alphabet. In these articles, Mereschkowsky extended his original hypothesis in 1905, and explained importance of symbiosis in biology, but the essential idea remained almost identical. Nevertheless, in the 1910 article, a curious figure illustrated multiple endosymbiotic origins of plastids with different pigments, each originated from a blue-green alga of the respective color. This was reproduced in Margulis and Bermudes (1985) along with a corresponding diagram of the authors themselves (Sect. 4.3.2).

Mereschkowsky thought that the nucleus of a eukaryotic cell originated from a bacterium *Micrococcus*. In his view, the first endosymbiosis with *Micrococcus* engendered a eukaryotic cell, which then incorporated a blue-green algal cell by endosymbiosis to create a chloroplast. It is not clear how the host cell before micrococcal endosymbiosis looked like, but Mereschkowsky (1910) used this to explain the differences in cell nucleus and cytoplasm. This idea was severely criticized by the supporters of Margulis, who denied entirely the value of the endosymbiotic theory of Mereschkowsky containing erroneous hypothesis. This criticism is still found in recent papers (Lazcano and Peretó 2017). I do not understand why these criticisms are not focused on the main part of Margulis theory (origin and evolution of mitotic mechanisms), which was also erroneous (see Chap. 4). Curiously, Mereschkowsky did not consider mitochondria as endosymbionts, maybe because the nature of mitochondria was still unclear in his time or later in the first half of the twentieth century, in which erroneous hypotheses on the endosymbiotic origin of mitochondria were repeatedly presented.

2.5 Symbiogenesis Notion

As we have seen, Mereschkowsky succeeded in assembling all kinds of knowledge of the time to finally propose an endosymbiotic hypothesis of plastid origin. We can imagine a Russian cultural background as I pointed out already, but we have to recognize the contribution of his own researches on diatoms and lichens. In addition, his strange personality also contributed to his brave act of proposing an unprecedented hypothesis. Based on his idea of endosymbiosis, he advanced to present a theory of "Symbiogenesis" (capitalized in German) in his 1910 article. This later became the central idea of Kozo-Polyanski in Russia, and was independently developed by Myer-Abich in Germany. Symbiogenesis is a notion that a new type of cell is produced by endosymbiosis of different kinds of cells being established as organelles. This will be a motif later incorporated into the Margulis theory on the origin of eukaryotic cells. A reason for supporting symbiogenesis at the time was that many researchers were frustrated with the inability of Darwinian theory of evolution to explain the emergence of novelty in biological evolution. It should be noted that a different usage of "symbiogenesis" existed in a context of creation of value by mutual cooperation, which was unrelated to endosymbiosis (Reinheimer 1915).

Chapter 3
Endosymbiotic Discourses Until the Mid-Twentieth Century

After the proposal of Mereschkowsky hypothesis, or in parallel to it, various hypotheses were presented including the idea of endosymbiosis. Sapp's book (1994) detailed historical arguments on these hypotheses. Khakhina (1979/1992) presented endosymbiotic discourses in Russia. Mitochondrial endosymbiosis was also proposed in the meantime, but was not supported by experimental evidence. In this chapter, I will present various discourses on symbiosis and endosymbiosis from the end of the nineteenth century until the mid-twentieth century.

3.1 Plastid Notion of Schimper

Andreas Franz Wilhelm Schimper (1856–1901) was a German botanist. His father Wilhelm Philip and his cousin were also researchers in related fields. Schimper is known for his studies on plant ecogeography and plant physiology. He published his chloroplast-related works, "Über die Entwicklung der Chlorophyllkörper und Farbkörper" (On development of chloroplasts and plastids. Schimper 1883) and "Untersuchungen über die Chlorophyllkörner und die ihnen homologen Gebilde" (Studies on chloroplasts and their homologous bodies. Schimper 1885). In these articles, Schimper stated that chloroplasts, leucoplasts, and amyloplasts among others are homologous structures that he called plastids (or chromatophores). Most importantly, he established that plastids are formed only by division of plastids.

> Die Chromatophoren gehen nämlich ausschliesslich aus anderen Chromatophoren hervor, sie entstehen nie durch Neubildung aus dem Zellplasma. (Namely, plastids are formed exclusively from other plastids, they never arise by new synthesis from cytoplasm.) (This phrase was entirely emphasized in the original. Schimper 1885, p. 4, my translation)

The Latin phrase "*omnis cellula e cellula*" (Every cell is derived from a cell), coined by a French biologist Raspail in 1825, was popularized by Rudolf Virchow (1821–1902) in 1855, and the correctness of this axiom was recognized in the time

N. Sato, *Endosymbiotic Theories of Organelles Revisited*,
https://doi.org/10.1007/978-981-15-1161-5_3

of Schimper (Morange 2016). Now Schimper established another axiom, "all chloroplasts from chloroplasts" or "all plastids from plastids."

Schimper described various properties of different types of plastids. He also described an idea on the origin of plastids in a footnote in the 1883 article.

> Solte es sich definitiv bestätigen, dass die Plastiden in den Eizellen nicht neu gebildet werden, so würde ihre Beziehung zu dem sie enthaltenden Organismus einigermaassen an eine Symbiose erinnern. Möglicherweise verdanken die grünen Pflanzen wirklich einer Vereinigung eines farblosen Organismus mit einem von Chlorophyll gleichmässig tingirten ihren Ursprung.
>
> (If it is definitely demonstrated that plastids are not newly synthesized in egg cells, the relationship between plastids and the cell containing them somehow reminds me symbiosis. Possibly, the origin of green plants could have been actually a fusion of a colorless organism with another containing chlorophylls uniformly.) (Schimper 1883, pp. 112–113, footnote 2, my translation)

Upon reading this, Mereschkowsky was so deeply impressed that he developed the idea in his way and he proposed the first endosymbiotic hypothesis of plastid origin. He jumped from the plastid continuity to the continuity of blue-green algae and plastids. In contrast, Schimper himself did not advance his studies on chloroplasts further, but he completed a large volume of biogeography in 1898.

3.2 Famintzin's Notion of Endosymbiotic Origin of Chloroplasts

Andrei Sergeivich Famintzin (1815–1918) was another proponent of endosymbiotic hypothesis of plastid origin. He was a professor of plant physiology in St. Petersburg Imperial University, which was located in the capital of the Russian Empire. He published many works on photosynthesis and chloroplasts, and was elected honorable president of the Russian Society of Plant Physiology in 1915. From the end of the nineteenth century, he started experiments on the isolation and culture of plant chloroplasts. He published a paper "Die Symbiose als Mittel der Synthese von Organismen" (Symbiosis as a method of organism synthesis) in 1907 in German biological journal *Biologisches Centralblatt*, and described a hypothesis of generation of a plant by symbiogenesis, in brief, an endosymbiotic hypothesis of plastid origin (Famintzin 1907). This is the same journal in which Mereschkowsky published his paper in 1905. Famintzin showed that lichens are composed of algae and fungi, and then tried to put forward a hypothesis on the generation of a new organism by synthesis of different organisms. He interpreted the experiment of chloroplast isolation as showing that chloroplasts proliferate independently of cell nucleus. He thought that chloroplasts could survive autonomously outside the cell, just as zoochlorellae and zooxanthelles do. Essentially, he developed the Schimper's idea of plastid continuity, and combined it with his studies on lichens to propose that free-living organisms can generate a new organism by symbiosis, as exemplified by plants. Therefore, his idea was not very different from that of Mereschkowsky.

Interestingly, Famintzin disliked Mereschkowsky, who did not have doctor's degree and was not a professor. He criticized Mereschkowsky's lack of knowledge and his confusion of zoochlorellae and plastids by citing Mereschkowsky's article in length (Famintzin 1907, p. 369). As discussed above, I believe that Mereschkowsky constructed his hypothesis by understanding the difference of zoochlorellae and plastids. This type of discussion might not be meaningful at the time when the distinction between eukaryotes and prokaryotes was still ambiguous, and we might not be able to understand the real significance of the dispute between Mereschkowsky and Famintzin. Famintzin also criticized as ill supported the similarity of plastids and blue-green algae stated by Mereschkowsky. Contrary to Famintzin's hostility against Mereschkowsky, the endosymbiotic hypothesis of plastid origin was diffused under the name "Mereschkowsky-Famintzin hypothesis" (see Sect. 3.6).

3.3 Various Hypotheses on the Endosymbiotic Origin of Mitochondria

Although the main subject of this book is endosymbiotic theories of plastid origin, it is pertinent to present some discourses on the mitochondrial origin as related information. The discussion dated from the late nineteenth century when the nature of mitochondria was still ambiguous.

3.3.1 Richard Altmann

Richard Altmann (1852–1900) was a famous German anatomist, and a professor in Leipzig University. He developed various staining (chromate, osmic acid, etc.) and examination methods to examine the cells of various organs, and described them in detail. He also gave the name "nucleic acids" to the acidic materials obtained by deproteination of Miescher's nuclein.

Altmann found various kinds of small granules within the cell, and described "Zellengranula" (cell granule) or "Bioblast" as well as oil droplets (Altmann 1889, 1894). Some researchers regard bioblast as mitochondrion, but the identity of bioblast is unclear as far as I read Altmann's writings. A more reasonable interpretation of his works suggests that he described general principles of cell organization, namely, that a cell is composed of bioblasts and protoplasm. Bioblast was described as a crystal of organized proteins or "Elementarorganism" (elementary organism), a basic unit proliferating within the protoplasm (Altmann 1894, Conclusion). Although such view appears interesting in the contemporary context, it was severely criticized and refused at the time.

3.3.2 Paul Portier

Paul Portier (1866–1962) was a French zoologist, who obtained a medical doctorate in 1897, and then a doctorate of sciences by studies on fungi in 1911. In 1901, he discovered the phenomenon of anaphylaxis during the oceanographic expedition with Charles Richet, who obtained the Nobel Prize in 1913 for this discovery. Portier published *Les Symbiotes* in 1918, in which he asserted similarity of mitochondria with bacteria (Portier 1918). He described cultivation of presumably symbiotic bacteria isolated from animal tissues, which he considered similar to mitochondria. He also stated that, in the digestive tubes of animals, bacteria play a role of incorporating nutrition and transferring it to animal cells. For this purpose, bacteria can enter the interior of animal cells and tissues:

> Si la nourriture est composée de substances réfractaires aux sucs digestifs de l'animal, le cryptogame attaque ces produits résistants au moyen de diastases appropriées, soit en végétant directement sur lui, soit en sécrétant ses ferments à travers la paroi des cellules qui le renferment. (…) Le microorganisme transformateur peut se développer dans le milieu extérieur, dans la lumière du tube digestif, dans les cellules qui bordent cette cavité, ou même dans la profondeur des organes de l'animal.
>
> (If the food is composed of substances that are not digested by the digestive juice of animal, the bacteria attack these resistant materials with appropriate enzymes, either directly incorporating them for their growth, or secreting these enzymes through the surrounding cell wall. (…) The microorganism having the ability to transform food may grow in the external medium, in the cavity of digestive tube, in the cells in the periphery of this cavity, or even in the interior of the animal organs.) (Portier 1918, pp. 200–201, my translation)

His postulate was criticized by researchers of the time, especially those of Pasteur Institute in Paris (Sapp 1994). Ivan Wallin was, however, influenced by Portier. Portier cited Altmann (1894) and Mereschkowsky (1910). He did not agree with Mereschkowsky who placed fungi, blue-green algae, and bacteria equally in amoeboplasma (=protoplasm) (Portier 1918, p. 100, footnote).

3.3.3 Ivan Wallin

Ivan Wallin (1883–1969) was an American anatomist and a professor in the Faculty of Medicine in Colorado University. Influenced by Portier, he proposed a hypothesis that mitochondria are symbiotic bacteria. Sapp (1994) wrote that Wallin used an English version of *Les Symbiotes* with translation mistakes to interpret in his own way what Portier wrote. Wallin published an article "On the nature of mitochondria. VII. The independent growth of mitochondria in culture media" (Wallin 1924), in which he was able to isolate mitochondria and culture them. This must have been an artifact of contaminating bacteria. In 1927, he published a book *Symbionticism and the Origin of Species* and affirmed that the origin of mitochondria was a microsymbiosis by bacteria (Wallin 1927). This term corresponds to endosymbiosis.

> It appears to the author that Symbionticism offers a rational explanation for many of the variations in the morphology and physiology of plants and animals. When these variations are of sufficient magnitude and permanence they constitute new species. The fact that mitochondria are universally present in the cells of all organisms higher than the bacteria, that mitochondria are bacterial in nature and that microsymbiosis can determine morphologic and physiologic changes in organs and cells, can lead to no conclusion other than that Symbionticism is a fundamental causative factor in the origin of species. (Wallin 1927, p. 114)

He considered that algae resulted from blue-green algal endosymbiosis, whereas flagella and cilia originated from endosymbiosis of bacteria having these motile apparatus.

> The ciliate and flagellate protozoa, apparently, acquired their special locomotor structures through symbiosis with ciliate and flagellate bacteria. We have previously mentioned that some investigators have claimed that cilia formation in the cells of higher organisms is associated with mitochondria. (Wallin 1927, p. 142)

Although this is an important book that Margulis cited later as a major source of her endosymbiosis theory, serious problems remained in its narratives, especially those on mitochondrial culture and its endosymbiotic origin. The spirochete hypothesis of Margulis had an origin in Wallin's writing as we see in the citation. Altmann (1889) and Merejkovsky (1920) were cited in this book. As I will explain later, no valid experimental supports were provided for the endosymbiotic hypothesis of mitochondrial origin until the discovery of mitochondrial DNA in 1963.

3.4 Symbiogenesis Hypothesis of Kozo-Polyanski

Boris Mikhailovich Kozo-Polyanski (1890–1957) was a Russian plant taxonomist and a professor in Voronezh. According to Khakhina (1979/1992), his activity on endosymbiotic hypothesis was limited to the 1920s, several years after he became a professor. He wrote only Russian articles and books. In 1924, he published *Symbiogenesis: A New Principle of Evolution* to emphasize that correct phylogeny should include hybrid formation, or more generally formation of new species by synthesis of different species (Kozo-Polyanski 1924/2010). At that time, Darwinism was not sufficient for explaining evolution, and a new principle of evolution had to be invented. Kozo-Polyanski also used lichens as a model of symbiosis and considered that symbiogenesis was an explanation that complemented Darwinian evolution. Citing Famintzin, he wrote that plants themselves are products of symbiogenesis, in which plastids originated from an endosymbiont.

The generalization of symbiogenesis was not supported by biologists. Kozo-Polyanski hypothesized that integration proceeds stepwise, from symbiosis to consortium, and finally to symbiogenesis. He explained evolutionary advantage of a holism involving production of a new whole by combining individual components. As a proof, he cited the zoogloea of terrestrial bacteria studied by Winogradsky as an example of bacterial consortium. Like Famintzin, he considered that the best proof of endosymbiotic origin of chloroplasts was a culture of isolated chloroplasts. Unlike Mereschkowsky, he did not consider blue-green algae as an origin of chloroplasts. He hypothesized a curious symbiosis, in which a chlorophyll-containing bacterium incorporated a colorless bacterium. He imagined that cell nucleus also originate from an endosymbiont. He found a similarity of mitochondria and symbiotic bacteria. Khakhina evaluated the book of Kozo-Polyanski (1924/2010) as a landmark in the history of symbiogenesis theory.

3.5 Origin of Life: Oparin and Haldane

The development of endosymbiogenesis theory, directly or indirectly, enabled emergence of a theory on the origin of life itself (Malaterre 2010). Many textbooks describe that a Russian biologist Alexander Ivanovich Oparin (1894–1980) published *The Origin of Life* in 1924 and proposed the coacervate theory (Oparin 1924). This Russian book was not known in Western World immediately, but the 1936 version was translated into English and distributed widely as a paperback (Oparin 1936/1953). The English translation of the original 1924 version is also available on the Internet. The book of Oparin began by historical discussion on the spontaneous generation theory. Louis Pasteur already established that life arises only from life in 1860s, but Oparin recognized that the origin problem is different from the everyday situation, and asserted that there is no other reasonable way to understand the origin of life than to hypothesize that the first life originate from nonliving matter in the early history of Earth. In fact, Pasteur also recognized the problem of the first life, which must be created from nonliving matter.

Oparin's theory was based on "chemical evolution," in which basic materials of life was produced by spontaneous chemical reactions from the materials present in the Space. These basic materials of life then assembled spontaneously to form a colloid-like protocell, which was converted to the first cell. This was a novel and original theory resulting from a synthesis of various lines of knowledge in different disciplines of the time. Morange (2016, p. 304) criticized, however: "Comme avant lui celui de protoplasma, le terme de colloïde semblait représenter en lui-même une explication et éviter de s'interroger plus avant sur le 'comment' des phénomènes décrits." (Like the term protoplasma before it, the term colloid seemed to represent itself an explanation, and to take away the question on the reason of the described phenomena). The mainstream scientists in physics and chemistry of the time considered that they would be able to understand life as an assembly of low-molecular weight substances, while ignoring the essential importance of polymers that construct biological materials. This was an expression of a hasty reductionism believing that life can be understood by materials science. In this sense, Oparin's theory was also reductionist, authorized by the communist government of the time, based on Marxist materialism. On the other hand, his method of searching the origin of life by combining earth sciences and biology was in line with that of Lynn Margulis half a century later.

A British biologist J. B. S. Haldane (1892–1964) also proposed a similar theory on the origin of life (Haldane 1929). This paper also began by the theory of spontaneous generation. He presented the theory of Félix D'Hérelle that the life originated from bacteriophages that had just been discovered. Haldane discussed the origin of life only briefly within several pages: after the cooling of the Earth, the action of ultraviolet light generated various organic substances and large molecules, which became "half-living things." He did not describe details of the process producing the initial cell. This is indeed a short essay without references. It is not an article comparable to the Oparin's book, which insightfully discussed the origin of life. However, in Western World, Haldane is believed to conceive chemical evolution independently of Oparin.

In the current biology, the idea of origin of life as imagined by Oparin and Haldane is accepted as reliable, but, if we check individual steps in the process from nonliving materials to the initial cell, there are a lot of premises, most of which have not been demonstrated by experiments (Malaterre 2010). Nevertheless, we have no other choice than to believe that the initial organism was generated by chemical evolution and self-organization. Because any of the pieces of processes that the researchers in chemical evolution pre-suppose does not occur easily, life can be realized only by a very rare chance that many events happen at an appropriate timing and sequence. I wonder what the researchers of the origin of life think honestly about this rare chance.

3.6 Pascher's Notion of Symbiosis in Algae

Adolf Pascher (1881–1945) was a German botanist/algologist and was a professor in Praha, which was a territory of Germany then. He published a long review on symbiosis, emphasizing "Syncyanose," a symbiosis of blue-green algae and other cells (Pascher 1929b). He used the terms Ectocyanose and Endocyanose for external and internal symbioses, respectively. In Part 1, he presented endosymbiosis of blue-green algae in flagellates, rhizopods, and algae. *Paulinella*, *Geosiphon*, Glaucophyta (*Gloeochaete* and *Glaucocystis*), and diatoms (*Rhizosolenia* and *Richelia*) were presented as examples.

Part II presented newly discovered examples of symbiosis, such as *Peliaina*. Pascher described isolation of cyanelles, as well as the cells of *Peliaina* that no longer contain cyanelles after uneven cell division. He also described Glaucophyta (*Cyanophora*), Cryptophyta (*Chroomonas*, *Cryptella*), and an amoeba containing blue-green algal cells within the interior. Next, he described a new example of symbiosis between a green alga (*Chlorella*) and a blue-green alga, such as *Cyanoptyche* and *Chalarodora*.

Part III proposed a systematic taxonomy of symbionts according to hosts and cyanelles. Part IV is devoted to the general theory of syncyanose, namely, (1) phototaxis, (2) relationship between animal nutrition and endosymbiosis, (3) relationship between starch accumulation in the host cell and the number of cyanelles, (4) relationship between the morphology of cyanelles and life style of endosymbiosis, and (5) variation of cyanelle numbers.

Pascher commented on the origin of plastids in Part V entitled: "Bemerkungen zur Chromatophorentheorie Famintzin-Mereschkowskys" (Remarks on the plastid theory of Famintzin and Mereschkowsky).

> Mereschkowsky und Famintzin haben bekanntlich die Hypothese aufgestellt, daß die Chromatophoren der Pflanzen ursprünglich nicht zelleigene Chromatophoren gewesen sind, sondern aufgenommene Algen, speziell Blaualgen darstellen, die völlig in die Funktion zelleigener Organe getreten sind. (As known, Mereschkowsky and Famintzin proposed a hypothesis that plant plastids have not been plastids inherent in the cell, but are incorporated algae, especially blue-green algae, that perform perfectly the function of cell organelles.) (Pascher 1929b, p. 454, my translation)
>
> Die prinzipielle Richtigkeit der Famintzin-Mereschkowskyschen Theorie hat, wie auch Geitler betont, als erwiesen zu gelten. Wir kennen eine immer mehr zunehmende Zahl von solchen Symbiosen, die der Famintzin-Mereschkowskyschen Annahme völlig entsprechen. Die Allgemeingültigkeit dieser Hypothese ist nicht erwiesen; ja ich meine nach allem, was wir über die Chromatophoren der Niedren wissen, wird sich die Hypothese in der von den Autoren geforderten Allgemeinheit nicht erweisen lassen. Durch die Famintzin-Merescbkowskysche Theorie wird eben nur eine der bestehenden Möglichkeiten erfaßt. (The basic correctness of the Famintzin-Mereschkowsky theory is, as Geitler emphasized, regarded as proven. We know increasing numbers of such symbioses that Famintzin-Mereschkowsky hypothesis perfectly matches. The general validity of this hypothesis is not proven. According to all what we know about plastids of lower organisms, I believe that the hypothesis will not be proven in a general form that the authors demanded. The Famintzin-Mereschkowsky theory is considered only one of the existing possibilities.) (*ibid*, p. 455, my translation)

Part VI is a summary, followed by descriptions of new species.

Pascher's article was published in 1929, only 8 years after the death of Mereschkowsky. We can clearly understand that his endosymbiotic hypothesis was well known in the academic world. Without knowing this fact, Lynn Margulis presented the hypothesis of Mereschkowsky and Famintzin half a century later as an original, but non-influential theory within the Russian academism that was not known in Western World:

> The studies of Famintsyn, Merezhkovsky, and Kozo-Polyansky laid the foundation for a burgeoning interest in symbiosis, (…) Their ideas were exceedingly slow in reaching the West, where the interest in symbiotic processes occupied a somewhat broader but more diffuse front. (Foreword by Margulis to the English version of Khakhina 1979/1992, p. xi, left column. Note that the spelling of Russian names was different from those used in this book.)

This is clearly not true, except for Kozo-Polyanski. Although known to many botanists in Western countries, the Mereschkowsky-Famintzin hypothesis encountered caution and criticism, and were not accepted in the first half of the twentieth century.

3.7 Buchner's Monumental Work on Symbiosis

It was a German zoologist and cytologist, Paul Buchner (1886–1978), who assembled all biological knowledge on symbiosis and endosymbiosis. He published a large volume *Endosymbiose der Tiere mit Pflanzlichen Mikroorganismen* in 1953 (Buchner 1953), in which he explained various examples of symbiosis. This book was translated into English in 1965 as *Endosymbiosis of animals with plant microorganisms*. The 1953 book included description of various examples (I added the number of examples in parenthesis), as indicated by the following list of chapters (my translation).

Discovery History and Extension of Endosymbiosis

1. Algal symbiosis
2. Symbiosis with fungi and bacteria
3. Wrong way of symbiosis study

Specific Topics

1. Symbiosis in animals that nourish on cellulose-containing substances, herbaceous plants, seeds, and similar parts (18 examples)
2. Symbiosis in animals living in vascular bundle of trees
3. Symbiosis in animals sucking tree juice (6 items including 19 examples)
4. Symbiosis in animals that suck vertebrate blood or eat scales (12 examples)
5. Symbiosis in insects living on various foods (3 examples)
6. Symbiosis in luminous animals (3 items including 9 examples)
7. Symbiosis in secreting organs (3 items including 6 examples)

General Discussion

1. Habitats of symbionts
2. Infection pathway
3. Embryo-level and postembryonic events
4. Exchange relationship between the host and its symbiont
5. Historical problems
6. Significance of symbiosis

What I think important is the reason why Buchner, after assembling a large amount of knowledge on various forms of symbiosis, was very cautious about interpretation of them. He strongly criticized a view that confused mitochondria and bacteria or chloroplasts and blue-green algae. He was worried about the crisis of integrity of eukaryotic cell, threatened by the idea of endosymbiotic origin of organelles.

> All diese kühne Hypothesen, durch welche die Endosymbiose zu einem fundamentalen, die Einheitlichkeit tierischer wie pflanzlicher Zellen sprengenden Prinzip erweitert wird, haben auf den Gang der Forschung keinerlei einfluß gehabt. (All these challenging hypotheses, by which endosymbiosis is extended to a fundamental principle that breaks up the integrity of

animal and plant cells, have had no influence on the progress of research.) (Buchner 1953, p. 80, my translation)

In the last part, he pointed out the relationship of animal immunity with symbiotic bacteria and fungi, and discussed the significance of symbiosis from the viewpoint of animal and human health.

Later, Lederberg tried to find a translator of Buchner's book, and the translation was finally realized in 1965. According to Sapp (1994), symbiosis studies of the time were separated into different fields, and a trans-disciplinary research organization on symbiosis was not established. The notion of symbiosis had negative connotation such as parasitism or social problems related to conflict, competition, and exploitation rather than mutual aid.

3.8 Lederberg's Endosymbiotic Hypotheses at the Dawn of Molecular Biology

Joshua Lederberg (1925–2008) discovered a bacterial sexual phenomenon, namely, conjugation of *Escherichia coli* cells, and subsequently formulated the notion of transduction, in which genetic characters are modified by introduction of genetic elements. In 1952, he demonstrated with Ester Lederberg that mutation is present before selection, using a replica plating method for bacteriophage. He was awarded the Nobel Prize for this important finding indicating that mutation is not a result of selection. On the other hand, he wrote a review article "Cell Genetics and Hereditary Symbiosis" by assembling preceding research results, and tried to explain virus and plasmid from the viewpoint of symbiosis (Lederberg 1952). I present section titles of his article below with some notes as necessary.

Genetic continuity of cytoplasmic organelles
Cytoplasmic inheritance
Chloroplasts: plastogenes and viruses

Lederberg argued that the continuity of chloroplasts was not clearly established morphologically, because distinction between proplastids and mitochondria was ambiguous at the time. Variegation was an important topic of research, but it was sometimes confused with viral infection. The term plastogene was introduced to describe genetic mutation of chloroplasts without identifying the nature of genes. Localization of plastogenes to chloroplasts was not identified unambiguously. It could even be viruses. Male sterility showed maternal inheritance, and could be due to mitochondrial mutation. A possibility of virus infection could not be excluded.

Cytoplasmic mutation in yeast

Petit mutation of yeast was known then, and its cause was localized to cytoplasm. Lederberg discussed that mitochondrial plasmid could account for the acclimation of yeast cells to galactose and its reversion by glucose addition.

Cytoplasmic inheritance in *Paramecium*

The macronucleus is produced from the micronucleus, and also regenerated from fragments of macronucleus. In this sense, Lederberg considered that both types of nuclei behave as independent plasmids. The cytoplasmic killer element kappa is eventually transferred to another cell upon conjugation. Kappa was considered a plasmid, which was maintained by the nuclear gene K. Kappa is involved in the production of a toxin called paramecin, but the relationship between kappa and paramecin seemed complicated. Antigenicity of cilia was also considered to inherit cytoplasmically, but involvement of the cell nucleus was not excluded.

Genetic transductions (transformations) in bacteria

Transformation was already known in *Pneumococci*. Lederberg proposed the term transduction to emphasize introduction of genetic materials. Transduction was also possible in *Bacillus subtilis*, and was accompanied by introduction of DNA, although the possibility of genetic activity of remaining protein was not excluded [Note that DNA was not identified as the genetic materials in 1952].

Endosymbiosis

Lederberg described that endosymbiosis of blue-green algae were well known, and listed examples such as *Geosiphon* (a phycomycete + *Nostoc*) and *Gloeochaete* (an apoplastidic chlorophyte + a cyanophyte). A particularly interesting example was the cyanelles of *Paulinella chromatophora. Peliaina cyanea* (chrysomonad or cryptomonad) has several cyanelles. He also listed symbiosis of zooxanthellae, zoochlorellae, and *Convoluta roscoffensis* (association of a flatworm and a chlorophyte). *Paramecium bursaria* was the best-known example of zoochlorella.

(Details are not shown for the sections that follow)

Insect symbioses

Other latent viruses, in plants and mammals

Lysogenic bacteria

Heterokaryosis, hyperparasitism, and sexuality in the fungi

Syntrophism and autotrophism

Self-reproduction and self-dependence

Evolution of the cell and organism

Currently, we know Lederberg only in the context of bacteriophage genetics, but he was an erudite of genetics, knowing wide-range of biological phenomena. As the article was published just before the double helix paper of 1953, we can understand the situation of the time that the relationship of DNA and cytoplasmic elements was still ambiguous. We can also find a curious chaos of organelles, cytoplasmic elements, and viruses in his article. Nevertheless, Lederberg affirmed that chloroplasts originated from endosymbiosis. Fifteen years later, Lynn Margulis described endosymbiotic origin of chloroplasts using the Lederberg article as an important source. This indicates that endosymbiotic theory became established in the discourses of some advocates in the mid-twentieth century.

3.9 Holobiosis Hypothesis of Myer-Abich

Adolf Myer-Abich (1893–1971) was a professor in Hamburg University in Germany and studied philosophy and theoretical biology. He emphasized holism and proposed a notion of holobiosis by extending the idea of symbiogenesis by endosymbiosis (Myer-Abich 1950). He identified the origin of this theory in the endosymbiosis notion first described by Schimper as well as the symbiogenesis notion of Mereschkowsky and Famintzin, but he criticized that the symbiosis that was described by Pascher and Buchner was far from holobiosis. He also criticized the opinion of Pascher (and Geitler) against the theory of Mereschkowsky that I cited in Sect. 3.6. He cited Portier and Wallin. Myer-Abich proposed that holobiosis was a driving force that engendered higher organisms. His writing was very close to the expression of Lynn Margulis in later times. Unfortunately, this was a German book, which remained neither recognized nor cited by Margulis. Myer-Abich presented the whole history of living beings from the origin of life. He also affirmed that plastids originated from holobiosis involving integration of blue-green algae in a eukaryotic cell. He formulated the holobiosis theory as a theoretical biologist, but did not affect later developments in endosymbiosis theory in modern biology.

3.10 Summary

In this chapter, we overviewed the discourses related to endosymbiotic hypotheses from the end of the nineteenth century until the mid-twentieth century. Most of these discourses were simply expression of ideas of researchers rather than established theories. It is true that the concepts of symbiogenesis and holobiosis were emphasized, without enough experimental data that supported them. Endosymbiotic hypotheses were supported by some researchers as theories that could explain large-scale evolution that complemented the Darwinian evolutionary theory, but were not recognized as general theories by most biologists. Until the rise of biological revolution in the latter half of the twentieth century, biology remained assembly of ad hoc discourses of leading scientists.

Chapter 4
Re-evaluation of the Initial Ideas of Lynn Margulis

Currently, any student studying biology knows the name of Lynn Margulis. The endosymbiotic theory is illustrated in a visually comprehensible way in any textbook, and nobody doubts about its truth. Even in the first half of the twentieth century, a limited number of biologists argued already about endosymbiotic hypotheses. However, in 1967, in which Margulis published her initial paper on the origin of mitosis, this was not a generally accepted theory. Why did Margulis propose her endosymbiotic theory as an original one? We have to re-examine her discourses in her initial writings.

4.1 Margulis Refuted Old-Fashioned Plant Phylogenetics

In her 1970 book *Origin of Eukaryotic Cells*, Margulis denounced the traditional botanical phylogenetics as "botanical myth," which envisaged an evolutionary path from blue-green algae to red algae, green algae, and finally to land plants. By these words, Margulis completely denied the traditional theory of plant evolution. Curiously, no particular criticism was raised against this challenging expression in her days. I suspect that the researchers of the time did not really understand the meaning of the expression. Nevertheless, the proposals of various forms of autogenous theories at the time had an intention of defending the botanical myth. Autogenous theory, also called continuous theory, implied that organelles were formed spontaneously from the components of the eukaryotic cells.

Marugulis tried to destroy the whole system of plant phylogenetics of the time, and this was not a minor correction. Margulis emphasized the essential distinction between prokaryotes and eukaryotes, and denied the phylogenetic scenario that blue-green algae became algae and plants by evolution. Figure 4.10a (left) shows the autogenous theory that Margulis tried to attack, or "botanical myth," in which ancestral blue-green algae generated flagellated algae, which engendered algae and plants.

N. Sato, *Endosymbiotic Theories of Organelles Revisited*,
https://doi.org/10.1007/978-981-15-1161-5_4

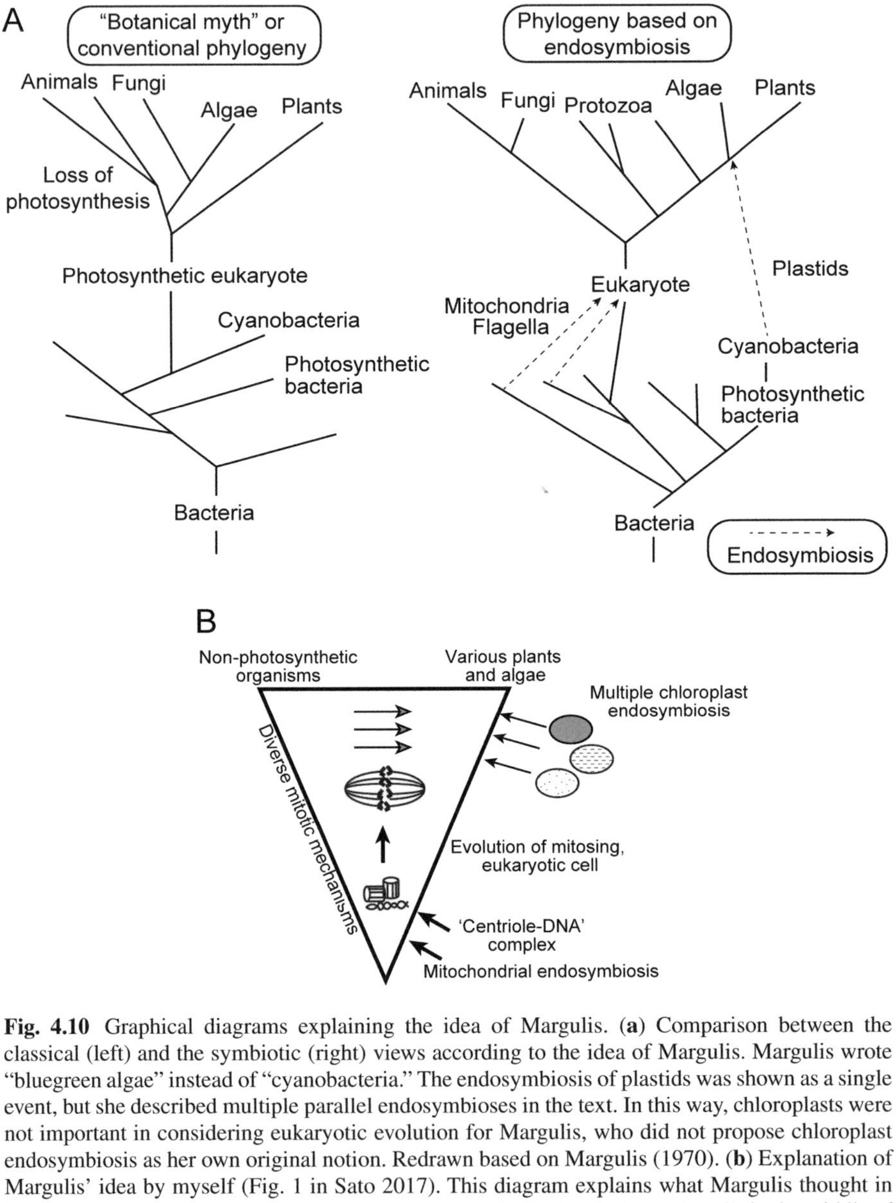

Fig. 4.10 Graphical diagrams explaining the idea of Margulis. (**a**) Comparison between the classical (left) and the symbiotic (right) views according to the idea of Margulis. Margulis wrote "bluegreen algae" instead of "cyanobacteria." The endosymbiosis of plastids was shown as a single event, but she described multiple parallel endosymbioses in the text. In this way, chloroplasts were not important in considering eukaryotic evolution for Margulis, who did not propose chloroplast endosymbiosis as her own original notion. Redrawn based on Margulis (1970). (**b**) Explanation of Margulis' idea by myself (Fig. 1 in Sato 2017). This diagram explains what Margulis thought in her early writings. She focused on the long evolution of mitosing, eukaryotic cells in the middle of the triangle. To establish a theory of eukaryotic evolution, she placed mitochondrial endosymbiosis as the earliest event, and multiple chloroplast endosymbioses as the last events that converted non-photosynthetic organisms to photosynthetic organisms. Reproduced with permission of the publisher

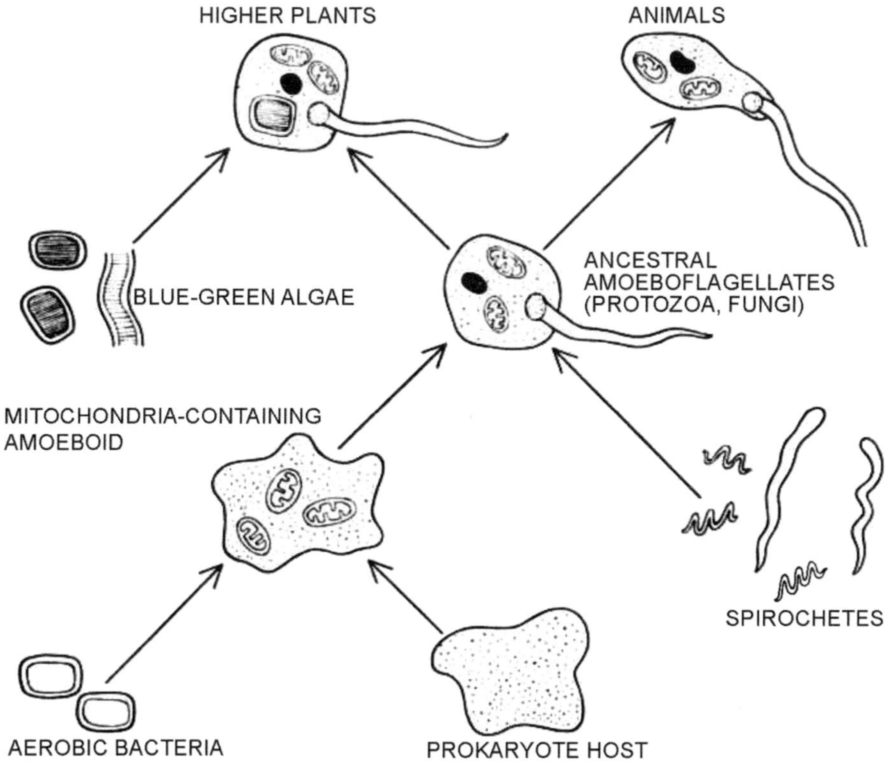

Fig. 4.11 Diagram showing the serial endosymbioses according to Margulis (1971). This figure begins at the bottom left. First, aerobic bacteria are ingested into a host cell to generate a eukaryotic cell. The aerobic bacteria became mitochondria. Next, spirochaetes became endosymbionts, thus producing an amoeboflagellates. This became animal cells according to the right arrow. On the other hand, the flagellates acquired blue-green algae to become plant cells having chloroplasts. Generation of multicellular organisms (plants and animals) was not shown explicitly. In the current knowledge, flagellated cells are known in only limited types of cells (sperms and epithelial cells) in animals. In plants, flagellated gametes are known in algae, bryophytes, pteridophytes, and some gymnosperms (*Ginkgo* and Cycads), but not in angiosperms. Plants and algae have two anterior flagella. These facts were known in the early twentieth century, but Margulis presented incorrect drawings, intentionally or unintentionally. Perhaps, the number of flagella was not a major focus of the researchers until Cavalier-Smith proposed a new classification of eukaryotes, Opisthokonta and Bikonta (Sect. 6.4). This figure was reproduced from Margulis (1971), but a similar figure was found in Margulis (1970). Reproduced with permission of the publisher

In addition, loss of photosynthesis in plants led to animals! This is curious, but remained probable at the time.

In contrast, Fig. 4.10a (right) illustrates that endosymbiosis resulted in mitochondria in a eukaryotic cell. Margulis formulated a simple pathway of serial endosymbioses of mitochondria, flagella, and then chloroplasts (Fig. 4.11). However, if we read carefully what Margulis wrote, the endosymbiotic process was not as simple as the one in Fig. 4.11, but as I depict in Fig. 4.10b, it involved a long, complicated process of eukaryotic evolution and diversification after the establishment of a eukaryotic cell, and completed by multiple parallel endosymbioses of chloroplasts. The main points of Margulis' idea included that invention of mitosis was essential in generating a eukaryotic cell, and that mitosis should be modeled according to the division and partition of basal bodies. These points were already stated in her 1967 paper. According to Margulis, the central event in the birth of eukaryotes was the introduction of (9+2) microtubular complex structure consisting of flagella, basal body, and its accompanying hypothetical DNA by endosymbiosis. She regarded it as an endosymbiosis of a spirochete, but she did not seem to stick to the origin, whether it was a spirochete or not, at least in 1967 or 1970. Nevertheless, she never abandoned the spirochete hypothesis throughout her life. Upon looking at Fig. 4.11, we have an impression that Margulis proposed the endosymbiotic origin of chloroplasts, but I suppose that what she intended to propose was that chloroplasts were dispensable in the major process of eukaryotic evolution, in other words, chloroplasts should be added after the completion of eukaryotic diversification.

Nevertheless, according to what was told about Margulis, her major achievement was her proposal of endosymbiotic origin of chloroplasts and mitochondria. I feel a gap between this evaluation and her proposals in her 1967 paper and 1970 book. We will find changes in the focus of her major statement from the initial stages until the time when her academic achievement was established. Namely, in early days, it was necessary to remove chloroplasts and mitochondria from the main pathway of eukaryotic evolution, and for this reason, she affirmed that these organelles were exogenous, or originated by endosymbiosis. In other words, the endosymbiotic theory of mitochondria and chloroplasts was not the main part of her initial proposal. Rather, these organelles were unimportant. However, she began to affirm later that it was herself that proposed the endosymbiotic theory of mitochondria and chloroplasts. This point will be discussed later.

In the following sections, we will see Margulis' arguments and logic in her 1967 paper (Sagan 1967), and then we will try to understand Margulis' original idea according to *Origin of Eukaryotic Cells* (Margulis 1970). How many researchers and students read the original writings of Margulis? Perhaps the situation might be different, but at least in Japan, I know only a few researchers who read the 1970 book, although it is still quite famous even among students. Nevertheless, I am sure that virtually no one understands the logic of Lynn Margulis in these writings of her early career. In the next section, I will try to explain the 1967 paper to the readers, emphasizing the main topic which the author affirmed as an expression of her originality. Note that not all references within the citations are listed in the reference list at the end of this book.

4.2 The 1967 Paper

4.2.1 Section 1 (Introduction)

The introduction first presented the distinction between prokaryotes and eukaryotes, citing Stanier et al. (1957/1963), and then showed the overall plan of the paper:

> This paper presents a theory of the origin of this discontinuity between eukaryotic (mitosing or "higher") and prokaryotic cells. Specifically, the mitochondria, the (9+2) basal bodies of the flagella, and the photosynthetic plastids *can all be considered* to have derived from free-living cells, and the *eukaryotic cell is the result of the evolution of ancient symbioses*. Although these ideas are not new [Mereschkowsky (1910) & Minchin (1915) in Wilson (1925), Wallin (1927), Lederberg (1952), Haldane (1954), Ris & Plaut (1962)], in this paper they have been synthesized in such a way as to be consistent with *recent data on the biochemistry and cytology of subcellular organelles*. In accord with both *the fossil record* and this theory, the lower eukaryotes (protozoans, eukaryotic algae, and fungi) can now be included on a single phylogenetic tree (Figure 1). In contrast to previous thought on the subject (Cronquist 1960; Dougherty & Allen 1960; Fritsch 1935), *many aspects of this theory are verifiable by modern techniques of molecular biology*.
>
> In defending the idea that the eukaryotic cell arose by a specific series of endosymbioses, a plausible scheme for the origin of mitosis itself emerges. ("Mitosis" is meant only in the classical sense; the analogous equal distribution of genes to daughter cells in prokaryotes is not relevant here.)
>
> The paper is organized into three parts. The first presents the theory of the origin of the eukaryotic cell. The second part is a compilation of the scientific literature in support of the sequences presented first. The last section suggests some experimental results which can be predicted from the theory. (Sagan 1967, p. 226. My emphasis. NB: not all citations are present in the reference list of this book)

This is the narrative of Lynn Margulis. First, she formulated the problem, namely, the fundamental discontinuity of eukaryotic and prokaryotic cells, in the preceding short paragraph and the first sentence of the citation. Then, she presented the main thesis that the eukaryotic cell is the result of ancient symbioses (my emphasis), although she presented various endosymbiotic events as just probable ("can all be considered"). All subsequent paragraphs were used to convince the readers by repeated explanations with variations like a symphony. This strategy is not limited to this paper, but is used consistently in her all writings all through her life, except her earliest papers reporting results of experimental studies (see, e.g., Sagan et al. 1965). A scientific paper, especially the one in biology, normally begins by setting the problem, and then presents methods to solve the problem and the results obtained with the methods. It ends with discussion and conclusion. An entirely different strategy of the narrative of Margulis, consisting of the main thesis followed by explanations, certainly embarrasses many scientists.

Another important point in this introductory paragraph was the statement that her theory was a renewal of the old ones (see the listed references), but supported by new scientific data (biochemistry, cytology, and fossil record). She also predicted that her theory would be verified by molecular biology developing at the time. In the dawn of biological revolution triggered by the rise of molecular biology, this novelty acted as a strong motive in convincing biologists to accept her major contribution to this field and her priority of the revised theory. This point will be discussed later again.

As stated in the second last paragraph of citation, one of the originality of her proposal resided in the origin of mitosis, which is clearly represented by the title of the paper "On the mitosing cells." The major contribution of Lynn Margulis, therefore, must have been her theory of mitosis, rather than endosymbiotic origin of chloroplasts and mitochondria. The relative weights of the three endosymbioses seemed to change with time in her writings. This will be another topic to be discussed later.

I regret to point out various minor typographical problems: First, Mereschkowsky was misspelled as Merechowsky. Second, style of reference citation was not consistent within a single paragraph. The 1967 paper has been brushed up during the process of 15 cycles of rejection and resubmission according to Margulis' later reflection (Margulis 1998). If so, how did the initial manuscript look like? Were these small mistakes inevitable for a young author of 29 years old without coauthor or supervisor? This will be another point of discussion.

4.2.2 Section 2: Hypothetical Origin of Eukaryotic Cells

This section described the main part of her narrative or theory, without presenting the reasons. It consisted of the following subsections (NB: To distinguish the section numbers in this book from those in the cited paper, the latters are presented within parentheses with "subsection." Uppercase was used in the original titles of subsections. I present a short summary for each items.):

(Subsection 2.1) THE EVOLUTION OF PROKARYOTIC CELLS IN THE REDUCING ATMOSPHERE

This part consisted of a short paragraph explaining that prokaryotic cells arose by natural selection under the reductive atmosphere from 4.5 to 2.7 Mya (billion years ago). The birth of the Earth was already correctly positioned at 4.5 Mya, which is now shifted to 4.6 Mya.

(Subsection 2.2) THE EVOLUTION OF PORPHYRIN SYNTHETIC PATHWAYS, PHOTOSYNTHESIS, AND RESPIRATION IN PROKARYOTIC CELLS

This part presented a then contemporary thought on the evolution of prokaryotes: "Early in the history of prokaryotic cells, nucleotide sequences (genes) which coded for porphyrin syntheses, evolved (p. 227. NB: punctuation is strange in the original)." She thought that metal-chelated porphyrin protected prokaryotic cells in the early Earth that was threatened by light-induced oxygen production, and she believed that natural selection accounted for this evolution. She added: "cells eventually evolved *containing mechanisms* to produce ATP (and the other nucleotides) utilizing visible solar energy absorbed by chlorophyll-like porphyrins. (NB: the italicized part is incomprehensible)" In this way, she believed that photosynthesis, ATP synthase, and cytochromes emerged. This was followed by heterotrophs: "Among autotrophic microbes which used porphyrins in anaerobic photosynthesis in the light, mutants which could use these same porphyrins for aerobic respiration in the absence of light were eventually *selected for* (my emphasis)." The atmosphere became oxidative as evidenced by the deposition of oxidized rocks in 2.4 Mya, as we know now. It should be noted here that the researchers of the time believed that oxygenic photosynthesis preceded aerobic respiration, and the narrative of Margulis was consistent with the general understanding of the time.

(Subsection 2.3) THE EVOLUTION OF EUKARYOTIC CELLS FROM PROKARYOTIC CELLS BY SYMBIOSIS

Margulis explained the birth of eukaryotic cells by an adaptation to aerobic environment: "the first step in the origin of eukaryotes from prokaryotes was related to survival in the new oxygen-containing atmosphere: an aerobic prokaryotic microbe (i.e., the protomitochondrion) was ingested into the cytoplasm of a heterotrophic anaerobe. This endosymbiosis became obligate and resulted in the evolution of the first aerobic amitotic amoeboid organisms (p. 228)." She believed that this evolution occurred about 1.2–0.6 Mya, but that the mechanism of mitosis did not exist at this stage. This endosymbiosis resulted, she thought, in establishment of common glucose oxidation pathway consisting of glycolysis and citrate cycle (she used the term "Krebs cycle"), and internal membranes (nuclear membrane and endoplasmic reticulum) consisting of phospholipids and sterols. In addition, "The greater amounts of energy available after the incorporation of the mitochondrion resulted in large cells with amoeboid and cyclotic movement (p. 229)." However, "in the absence of any efficient mechanism to insure the equal distribution of newly synthesized DNA, the size of the amoeboids host cell must have been limited." "The problem of equal segregation of the daughter linkage groups remained."

Now, her original narrative on the evolution of mitosis was disclosed.

(Subsection 2.4) THE EVOLUTION OF MITOSIS IN AMOEBOFLAGELLATES

Margulis showed various types of mitosis in a number of figures, and tried to reveal the evolutionary history of eukaryotic cells. Mitosis began by an endosymbiosis:

> … it is hypothesized that some ancestral amoeboid acquired a motile parasite, perhaps spirochaete-like, by ingestion. The genes of the parasite, of course, coded for its characteristic morphology, (9+2) fibrils in cross section. Amoeboid host cells which retained their endosymbionts had the immense *selective advantage of motility* long before mitosis evolved; they could actively pursue their food.
>
> By hypothesis, the motile prokaryotic endosymbiont itself is the ancestor of the complex flagellum of eukaryotic cells. The replicating nucleic acid of the endosymbiont genes (which determines its characteristic (9+2) substructure) was eventually utilized to form the chromosomal centromeres and centrioles of eukaryotic eumitosis and to distribute newly synthesized host nuclear chromatin to host daughters." (pp. 230–231. My emphasis)

This was the core of the initial idea of Lynn Margulis! It is interesting to note that Margulis always used the notion of natural selection and selective advantage to explain the evolutionary scenario that she tried to demonstrate. She announced that mitosis is a new and pivotal criterion of classification in eukaryotic microorganisms:

"The groups were formed to include as much data as possible on the eumitosis of lower eukaryotes in a way which is considered to be consistent with the general classification of these organisms, *exclusive of their photosynthetic plastids* (p. 231. My emphasis)." This is a denial of the "botanical myth" (see Sect. 4.1) *by removing chloroplasts from the criteria of classification*. This appears to contradict with her affirmation of the importance of photosynthesis in the origin of prokaryotes (subsection 2.2). Once established, photosynthesis did not appear to play roles in natural selection in her mind.

(Subsection 2.5) STEPS IN THE EVOLUTION OF EUMITOSIS

This part occupied the major space (pp. 233–249, about a half of the whole article) in the 1967 paper, and undoubtedly, the main focus of the entire narrative of Margulis. Unfortunately, however, this part was very difficult to understand. Perhaps nobody ever understood her theory on the evolution of mitosis. I try to decipher her thinking at least partially. An important notion was "(9+2) homologue":

> the term "(9+2) homologue" is used here in its most general sense; *it implies the genetic system* that codes for the development of flagella, cilia, "division centers", centrioles, or any other of these homologous organelles. (p. 233. My emphasis)

In her evolutionary history, the "(9+2) homologue" changed its number, form, and function to enable different modes of mitosis. Curiously, she thought that it was a complex of genetic materials and a structure consisting of microtubules that were visualized as the "(9+2) homologue" in early 1960s. She described six steps in the evolution of mitosis. We will discuss this important notion later (Sect. 4.3.1).

(Subsection 2.6) THE EVOLUTION OF EUKARYOTIC PLANTS FROM VARIOUS LINES OF PROTOZOANS WHICH ACQUIRED SYMBIOTIC PROKARYOTIC ALGAE

Margulis considered that chloroplasts and photosynthesis were later acquisitions in eukaryotes:

> As outlined above, the evolution of mitosis is assumed to have occurred millions of years after the evolution of photosynthesis. This hypothetical origin of eukaryotes is completely incompatible with the notion that a simple phytoflagellate is the ancestor to extant higher cells—eukaryotic plant cells did not evolve oxygen-eliminating photosynthesis which later "packaged" into membrane-bounded plastids; they acquired it by symbiosis. The diversity of cell structure and the life cycle in lower eukaryotic algae imply that different photosynthetic prokaryotes (protoplastids) were ingested by heterotrophic protozoans at various times during the evolution of eumitosis. … This implies that in eukaryotic plants the plastids are homologous to blue-green algae; the non-chloroplast part of plants is directly homologous to eukaryotic heterotrophs. (pp. 244 and 247)

Margulis used "oxygen-eliminating photosynthesis" instead of "oxygen-evolving photosynthesis" that we currently use. The central notion stated in this citation was the acquisition of chloroplasts after finishing all eukaryotic evolution. Exclusion of chloroplasts was essential in endosymbiotic theory of Margulis (subsection 2.4). Chloroplasts (and mitochondria) were considered endosymbionts because they are not involved in eukaryotic evolution, which must be explained by the diversification of mechanism of mitosis which was her principal interest.

4.2.3 *Section 3: Evidence from the Literature*

This section had to be a collection of evidence from the literature, but apparently Margulis repeated her thoughts and hypotheses as in the preceding section. Basically, she accepted characterization of mitochondria and chloroplasts detailed in various review articles. She clearly stated to exclude photosynthesis from the taxonomical criteria in subsection 3.1. In this case, chloroplast endosymbiosis was nothing other than a means for constructing eukaryotic phylogeny rather than an important principle of its own.

(Subsection 3.1) ON CRITERIA FOR A NATURAL MICROBIAL PHYLOGENY

Margulis discussed problems in microbial taxonomy, and found a good criterion using molecular data:

> In determining the relationship of two microbes—that is, the amount of time elapsed since they diverged from a common ancestor—we may ask: how many homologous base pair sequences in DNA do they share? The number of mutational steps which occurred to produce one from the other is related to the number of generations elapsed since the two populations diverged. (p. 249)

This was exactly the principle of molecular phylogeny or the notion of molecular clock that would be utilized later. But in practice, it was not easy to apply. She correctly pointed out a good criterion:

> the point at which metabolic pathways diverge in two otherwise similar microbes may help determine the elapsed time since the two microbes themselves diverged from a common ancestor. (p. 249)

She presented some examples of this method, such as the taxonomy of lower plants based on photosynthetic pigments, but at the same time pointed out problems associated with taxonomy using plastid characteristics disregarding cellular structural characteristics.

(Subsection 3.2) SOLUTION OF SOME TAXONOMIC PROBLEMS

Margulis tried to explain discrepancy in taxonomy of microbes by separating the plastids and the remaining cell body:

> Consistent with the above-mentioned hypothesis is the theory that algae, with chrysophysean photosynthetic characteristics, evolved millions of years before from photosynthetic prokaryotes and were acquired symbiotically in heliozoan, rhizopodal, and sarkodinal species, and that these prokaryotic algae became the obligately symbiotic plastids in the various protozoans. (p. 252)

She added examples of euglenids and chloromonads, which presumably acquired green prokaryotic algae. These are now known as examples of secondary endosymbiosis (Sect. 6.4.2). She also listed some examples of a different kind, namely, similar host cells plus different types of plastids:

> There is really no other reasonable explanation for the evolution of certain natural groups of highly specialized flagellates which are found to contain distinct types of plastids with distinct pigments and storage products [e.g. Cryptomonadina (family): Rhodomonas (red), Cryptochrysis (chrysophysean), Chilomonas (colorless)]. (p. 252)

Cryptophytes are now known as a good example of secondary endosymbiosis, but the differences in plastid color, in this case, is not a result of independent endosymbiosis of differently colored prokaryotes. Rather, this is just a result of evolution within Cryptophyta. Margulis presented also an example of secondary loss of plastids, such as *Paramecium bursaria* and *Euglena gracilis*. Nevertheless, these two are now considered fundamentally different: *P. bursaria* is a real symbiotic system that can lose symbiotic algae, whereas *E. gracilis* cannot lose plastids, but retains degenerate plastids even in the dark. She formulated: "Thus, in general, a greater mutual dependency probably reflects a longer host-symbiont association (p. 253)." All these examples supported, she believed, her proposal that microbial taxonomy should be performed based on the mechanism of mitosis rather than photosynthetic traits.

(Subsection 3.3) GENERAL PROPERTIES OF SYMBIOSIS

This part described general properties of symbiosis, which are entirely hypothetical discussion. Nevertheless, we can understand the notion of symbiosis at the time, which was not very different from that of Mereschkowsky and Lederberg. Margulis listed properties of symbiosis, because she intended to use them as criteria of symbiosis, although this is logically erroneous in a strict sense. If an entity A possesses a certain property B, then, can we deduce that all the entities having the property B are classified as A? Certainly, not.

Nevertheless, Margulis asked: "What are the general criteria of organelles derived by symbioses?", and listed six "criteria" to judge whether an organelle was a symbiont or not. More exactly, these were hypothetical properties of an organelle that would have been acquired by endosymbiosis, but not criteria (pp. 254–255):

1. A symbiont originated as a free-living cell and therefore must have once been able to replicate its own DNA on its own protein synthesizing machinery. … Upon entry into a host, such a symbiont may lose from none to all of its synthetic capabilities *except the ability to replicate its own DNA* and synthesize complementary mRNA from that DNA—the *sine qua non* of any organism. Hence, if any of these organelles originated as symbionts, their characteristic specific DNA must be present in the host at every stage of the host life cycle (Emphasis and italics by Margulis).
2. If an organism or organelle has been acquired by symbiosis, it will be retained intracellularly in its host if, and only if, there exists some mechanism to insure that each daughter cell of the host receives at least one copy of the symbiont genome at each division.
3. If a cellular organelle is acquired by symbiosis, there should be no organism containing intermediate intracellular stages of the organelle.
4. If the symbiont is lost, all metabolic characteristics coded for on the symbiont genome must be lost together.
5. Since any intracellular symbiont must have its own genes, a correlation can be made between genetic traits conferred on the host by the symbiont and the morphological presence of the symbiont.
6. If an organelle originated as a free-living cell, it is possible that naturally occurring counterparts still can be found among extant organisms.

In short, genetic independence was taken as a sign of endosymbiotic origin of an organelle. This was identical to the notion of Lederberg (1952). I have to comment on the first criterion. The replication enzyme of organellar DNA is now known as "POP," which originated from a virus but not endosymbiont (Sect. 7.4).

(Subsection 3.4) ANCIENT ANAEROB1OSIS AND MICROBIAL PHOTOSYNTHESIS

This part described ancient history of the Earth related to the origin of life, citing the articles of her first husband Carl Sagan:

> All of the evidence is consistent with the widely accepted hypothesis that life arose under reducing conditions of the primitive atmosphere (Sagan 1965b). … The strong ultraviolet light absorption capabilities of the major nucleic acid components, of ATP, and of amino acids and peptides may have led to their local accumulation in the absence of replicating cellular systems upon which natural selection acted, as early as 4 to 5 billion years ago (Sagan 1961). … Correct identification of the fossils implies the origin of photosynthetic metabolism which uses the visible portions of the spectrum, occurred in cellular systems earlier than 2.1 billion years ago or even 3.1×10^9 years ago. … By hypothesis, the incidental fact that these antimutagenic tetrapyrole compounds were strong visible light absorbers was later put to advantage in the evolution of chlorophyll-mediated microbial photosynthesis (Sagan 1961). … It is possible that mutations causing the porphyrins to be used in cellular photoproduction of ATP originally occurred in microbes which already contained the CO_2 fixing "dark reactions". This could have resulted in the evolution of the fundamental photosynthetic mechanism, providing a new source of ATP, produced by visible light absorbed by porphyrins. In some organisms, chemotrophic or direct heterotrophic methods of obtaining ATP presumably were then replaced by the visible light-chlorophyll mediated methods characteristic of all microbial photosynthesizers. … The fact that oxidized limonite red-beds are as old as 2.5×10^9 years is strong evidence that mutations for oxygen eliminating photosynthesis occurred in some microbial autotrophs well before that time (pp. 257–259. NB: the two citations of C. Sagan are not included in the reference list of this book).

This was a scenario of life emergence popular in the 1960s. Life (anaerobes) emerged under the reductive atmosphere, and then photosynthesis (by cyanobacteria) started before aerobes. Currently, these two points are not accepted as correct. Nevertheless, we are surprised to see the extensive knowledge of Margulis on Earth science, which she certainly owed entirely to Carl Sagan.

I have to comment also on the role of ATP. In the 1950s, ATP was recognized as a "molecular unit of currency" of intracellular energy transfer, namely, ATP is involved in both metabolism and motility. From the viewpoint of driving force of life, reducing power (NADH and NADPH) is more important. The combination of the reducing power and oxygen provides the cells with the power of life activity (Sato 2012). Biologists in the 1960s knew photosynthetic production of carbohydrate and oxygen, but did not recognize the essential role of photosynthesis in producing reductants and oxidants.

(Subsection 3.5) ATMOSPHERE OXYGEN AND THE ORIGIN OF AEROBES

This part described emergence of aerobic microbes, which was made possible by accumulation of oxygen in the atmosphere caused by blue-green algal photosynthesis. She inferred:

> Under such conditions, constant selection pressure must have been placed on any microbes capable of tolerating and utilizing oxygen. ... Such mutations leading to the aerobic breakdown of carbohydrates presumably occurred in some photoautotrophs too, evolving the ancestors of blue-green algae. ... from 2.1 to 0.6×10^9 years ago the prokaryotic photosynthesizers eliminated more and more oxygen into the atmosphere. In response to this, different types of microbes evolved (including the proto-mitochondrion and prokaryotic aerobic algae), capable of coping with the increasing abundance of oxygen. (p. 260)

Margulis denied the continuity of blue-green algae and eukaryotic algae:

> The monophyletic evolution of eukaryotic algae from blue-greens is hard to reconcile with many other facts. Blue-greens are typical prokaryotes: they lack a membrane bounded nucleus, mitochondria, flagella, mitosis, sex, simultaneity of respiration and photosynthesis, and, in general, they are smaller than other algae. Furthermore, no intermediate organisms between blue-green and green algae have ever been found, even though they have been extensively sought. (p. 260)

Then she asked a central question, and presented her answer:

> Why then are the photosynthetic mechanisms in prokaryotic blue-green algae and eukaryotic green algae so similar? If the prokaryotic blue-greens were not ancestral to the eukaryotic algae and the higher plants, what organisms were?
>
> This apparent paradox can be resolved by recognizing the validity of the thesis presented in this paper, namely, that the evolution of photosynthesis preceded the evolution of the eukaryotic cell by millions of years, and that green plant oxygen-eliminating photosynthesis—so characteristic of both blue-green algae and "chlorophytes"—evolved in prokaryotes and was later acquired symbiotically by various eukaryotes. (pp. 261–262)

(Subsection 3.6) THE REPRODUCING MITOCHONDRION

In this part, Margulis summarized evidence that mitochondria can be regarded as endosymbionts, by citing Gibor and Granick (1964), Wallin (1927), and Wilson (1925). All evidence including self-duplication with specific DNA and RNA was obtained from the review Gibor and Granick (1964).

(Subsection 3.7) THE REPRODUCING (9+2) HOMOLOGUES AND THEIR RELATIONSHIP TO THE NUCLEUS

This part started with the observation of a strange flagellate *Trichonympha* reported by Cleveland in 1956, focusing on the function of centrioles in mitosis and its relationship with flagella. Margulis was ambitious and eloquent:

> it is unnecessary to explain how the self-replicating body differentiated from the nucleus before it was selected as a division center. The alternative assumption that a pre-existing replicating endosymbiont, the basal body of the flagellum, was utilized in a new role, is consistent with traditional belief—that evolution is opportunistic and not foresighted. (p. 266)

She cited observation of the division of basal body (called blepharoplast) in *Trypanoplasma* (Cytobia, an excavate belonging to the Kinetoplastida) and described it as a second mitosis. She also cited articles showing that the kinetoplast located at the base of flagella in another kinetoplastid *Leishmania* contained DNA. Kinetoplast is now known as a special form of mitochondrion, and the presence of DNA in kinetoplast is no longer evidence for the presence of DNA in basal body. She argued "genetic autonomy" of centrioles. Then, she put forward the spirochete hypothesis:

> For various reasons, spirochaete or spirochaete-like organisms have been suggested as likely candidates for free-living counterparts of the motile (9+2) endosymbiont which later differentiated into the flagellum, centrioles, and chromosomal centromeres. (p. 267)

A very strange organism *Myxotrixa paradoxa* is mentioned in the footnote as a note added in proof. It will be described in detail as an organism using spirochetes as flagella-like motor apparatus in the 1970 book.

(Subsection 3.8) THE REPRODUCING CHLOROPLAST
Margulis confidently stated: "Presumably because it was the least ancient of the symbioses, the exogenous origin of the photosynthetic plastid is the easiest to defend (Ris & Plaut 1962, Echlin 1966) (p. 268)." She cited her own work on DNA localization in *Euglena* chloroplasts. More than ten articles were cited in this part, and various pieces of evidence were checked with her criteria (see subsection 3.3) one by one. The logic was clearer compared with the previous subsections.

4.2.4 Section 4: Some Predictions

This uncommon section listed various predictions, assuming the validity of the theory of Margulis. She began by mitosis:

> if the theory presented in this paper is correct, all eukaryotes should ultimately be classified completely, correctly, and consistently according to their position in the origin of mitosis. (p. 270)
>
> Identification of (9+2) homologue-specific DNA and a complete characterization of its RNA and of its limited biochemical functions should eventually be possible. This is true of mitochondrial and plastid DNA as well.
>
> If these organelles did indeed originate as free-living microbes, our advancing technology should eventually allow us to supply all growth factors requisite for *in vivo* replication of all three of them—the *coup de grâce* to genetic autonomy (p. 270. Italics by Margulis. "Coup de grâce" is misused here in the meaning of "decisive factor".)

She added many more predictions:

> It is likely that biochemical pathways entirely unique to eukaryotes (e.g. steroid synthesis) should have component parts coded for by more than one organellar genome. (p. 270)
>
> the following organisms should have evolved: a free-living complex flagellar counterpart; a free-living mitochondrion counterpart; and a heterotrophic prokaryote capable of ingesting cells. (p. 270)
>
> If the theory is correct all eukaryotic cells must be seen as multi-genomed systems. ... All eukaryotes must contain at least three specific types of DNA: nuclear, mitochondrial, and (9+2) homologue. An additional DNA that is associated with the chloroplasts must be found in all eukaryotic plants. The plastids ... should have homologous plastid-specific nucleic acid. At the same time, those organisms which demonstrate "host homologies" ... should share homologous nuclear, and not necessarily plastid, DNA's. This is quite analogous to the presence of two distinct DNA bands of nearly equal size found in CsCl density gradient runs on DNA isolated from *Paramecium bursaria* (Sagan, L., 1964, unpublished data) (p. 271. NB: the citation is not included in the reference list of this book)

As already mentioned, she presented an interesting idea of molecular clock:

> When we know the precise metabolic pathways and their genetic basis, we will be able to calculate directly the number of generations elapsed since each organism evolved from a common ancestor—from the number of mutational sites at which they differ. (p. 271)

She dared to present unnecessary researches:

> Some searches will continue to be futile: for example, attempts to find eumitosis in all eukaryotes (if found, it will clearly be analogous, rather than homologous, to eumitosis in higher eukaryotes, cf. reports of sexuality in *Noctiluca*, a dinoflagellate); "missing links" in the origin of the ancestral phytoflagellate, such as organisms containing chloroplasts but no mitochondria; eumitotic organisms with bacterial flagella, or eumitotic fossils dating from anaerobic times. (p. 271)
>
> As in the past, future attempts to relate the various classes of algae directly to each other will be futile. For example, the ancestor of the relatively recent and isolated algae the *Bacillaracea* (diatoms), might be better sought among primitive metazoans than the chrysophytes or dinoflagellates. A simple flagellated organism with characteristic rhodophyte pigments probably never existed. A flagellated ancestor to the true molds and premitotic amoebae probably did not either. (p. 272)

This idea of the lack of phylogenetic relationship between various algae turned out to be false in the light of current phylogeny, namely, all algae belong to Bikonta. She closed her work by acknowledging J. D. Bernal (biochemist), C. Sagan (astronomer), and R. A. Lewin (phytologist) among others.

4.2.5 Discussion on the 1967 Paper

We have to discuss several important features of the 1967 paper. The first point is the abundance of evolutionary arguments. Curiously, Margulis considered that organisms are easily adapted to the environment by mutations and natural selection. She affirmed some special evolutionary changes that cannot be explained without symbiosis of prokaryotes (see Sect. 4.2.3 subsection 3.5), while new forms of prokaryotes can be produced by environmental adaptation. She repeatedly used "mutation" and "selected for" in the paper. This seems to contrast with her later attitude emphasizing the importance of symbiosis rather than Darwinian evolution:

> I felt that the field of population genetics, with its insistence on overly abstract neo-Darwinian concepts such as "mutational load," "fitness," and "coefficients of selection," taught more of a religion than a description of the rules by which real organisms passed on their genes and evolved. (Margulis 1998, p. 18)

The 1967 paper was a strange amalgam of "a just so story" of Darwinian evolution and a notion of symbiogenesis that was not consistent with Darwinian mode of evolution.

Second, combining amoebas, algae, and fungi into a single large clade was an original idea of Margulis that was based on her conviction that all eukaryotes should be classified according to the mode of mitosis, or more exactly according to the evolution of the mechanism of mitosis. Her original narrative explained how various eukaryotes were formed. She believed indeed that all eukaryotes can be or must be classified by the mode of mitosis.

In contrast with our contemporary understanding that algae (Archaeplastida) and plants are monophyletic, Margulis emphasized the differences in mitosis of algae and plants. She considered that red and green algae independently acquired plastids after their mitotic mechanisms were established. This is radically different from the current notion of plant evolution involving endosymbiotic theory of chloroplast origin. Curiously, despite marked differences in mitosis in green algae and plants, Margulis supposed a single endosymbiotic event for the green lineage. This could have been an inconsistency in evaluating the importance of mitotic mechanisms in evolution, but her reasoning remained unclear.

Mereschkowsky also proposed that various algae of different colors independently acquired plastids corresponding to their colors. The idea of multiple independent acquisitions of plastids was, therefore, shared by Margulis and Mereschkowsky. However, Mereschkowsky did not consider that the host cells were diversified before plastid endosymbiosis. He just hypothesized that different algae were produced by endosymbiosis of different types of blue-green algae. In spite of her belief of multiple algal origins, Margulis often showed simplified diagrams (Figs. 4.10a and 4.11) presenting only a single process for endosymbiosis of each organelle, which must have misled many readers. In a recent article, Martin (2017, p. 5) pointed out this inconsistency: "Today, everybody would probably agree with the cyanobacterial part, but not with the multiple origins."

Nevertheless, most algae that were targets of taxonomical re-examination in the 1967 paper were later identified as products of secondary endosymbiosis (Sect. 6.4), which occurred independently many times. In this sense, the idea of Margulis was partially right. She was wrong in that she considered multiple independent endosymbiosis of various types of blue-green algae.

4.3 The 1970 Book: Origin of Eukaryotic Cells

We now examine the ideas of Margulis in more detail according to her main work published in 1970, *Origin of Eukaryotic Cells*. We will focus on important points, avoiding duplication with the 1967 paper.

4.3.1 (9+2) Homologues and the Origin of Mitosis

Figure 4.12 shows electron micrographs of flagella and basal body in *Chlamydomonas reinhardtii*. The readers are certainly interested in the presence of a highly symmetric structure within a small flagellate, which must have also impressed researchers in the 1960s. In the cross section of a flagellum, we find two central microtubules, which are surrounded by nine pairs of microtubules. This has been called "(9+2) structure." Curiously, basal body has nine triplets of microtubules, but no central microtubules. This has been called "(9+0) structure." According to microtubule researchers, the 1960s were the time in which researchers began to recognize the generality of microtubules in eukaryotic cells (Brinkley 1997; Boisy et al. 2016). This must have been an impetus that prompted Margulis to conceive a hypothesis on the origin of mitosis. We can easily understand the situation of Margulis then, but her hypothesis was not easy to understand.

Importance of centriole in eukaryotic mitosis is well known, but it is not the origin of spindle microtubules. The assembly of spindle microtubules starts from a microtubule organizing center (MTOC). The notion of MTOC was proposed by

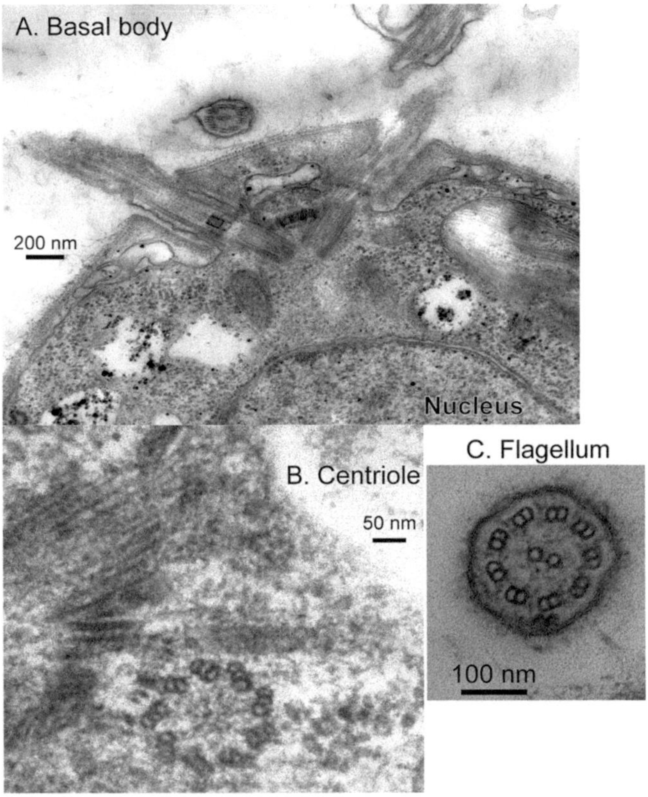

Fig. 4.12 Electron micrographs showing the flagella and basal body of *Chlamydomonas reinhardtii*. (**a**) Basal part of flagella. The nucleus is located in the proximity. It is known that many microtubules extend from the basal body located at the crossing of the basal parts of the two flagella and cover the surface of the nucleus. A mitochondrion and a part of chloroplast are seen on the left and the right, respectively. Some structure is found between the basal body and the nucleus. A hinge is found to connect the two basal parts of flagella. A cross section of another flagellum is also seen above the cell surface. (**b**) Basal body. Basal body is a derived form of centriole located at the base of a flagellum. Nine sets of microtubule triplets ("9+0" structure) are seen in cross section. A longitudinal section of another basal body is seen above. Thus, two basal bodies are arranged at the right angle. Each flagellum extends from each basal body (see panel **a**). (**c**) Cross section of a flagellum. Peripheral nine microtubule doublets and two central microtubules are seen. They are packed within a plasma membrane. Original figure

Pickett-Heaps (1969), who opposed Margulis later on the role of centriole. Currently, centriole is a part of centrosome that also includes MTOC. Margulis considered that exogenously acquired (9+2) structure contained DNA, and this complex that she believed centriole has an ability to divide by itself. In her view, the mechanism of binding of DNA and (9+2) structure (=microtubules) was somehow copied to the mechanism of binding of chromosomal DNA (=centromere) and spindle microtubules, which realized mitotic mechanism in eukaryotic cells (Fig. 4.10b). In other words, she thought that an ancestor of eukaryotic cell acquired the mechanism of DNA partitioning. By carefully comparing the processes of protist cell divisions, she noticed the presence of a similar mechanism of cell division in different organisms with or without chloroplasts. This convinced her that mechanism of cell division evolved first, and then various algae were created by acquisition of chloroplasts. She classified the following six combinations of centriole functions and mitotic mechanisms, which represented different steps of eukaryotic evolution. Similar descriptions are found in Sagan (1967, pp. 233–244) and Margulis (1970, pp. 263–271).

Step 1: (9+2) Homologue as basal body for flagella only
Step 2: (9+2) Homologue incorporated into the nucleus for division center, no flagella
Step 3: (9+2) Homologue used both as intranuclear division center and as flagellar basal body
Step 4: (9+2) Homologues used as basal bodies to flagella and other (9+2) homologues permanently differentiated as intranuclear division centers
Step 5: (9+2) Homologue used both as basal body to flagella and as extranuclear "division center"
Step 6: (9+2) Homologues used as basal bodies to flagella and other (9+2) homologues permanently differentiated as extranuclear division centers (centrioles)

Figure 4.13 depicts her model of Step 5, in which basal body is duplicated, and a new copy acts in mitosis. In short, a model of binding of a microtubule-based structure (string) and DNA (dot) was a prerequisite for Margulis. She considered another copy of endosymbiont DNA (the hypothetical DNA present in the "(9+2)" structure) was moved to chromosomes and became a centromere. Her model did not illustrate chromosomes or spindle explicitly, which made this model incomprehensible. Centriole was an original entity, and then a homolog appeared as kinetochore that binds microtubules and chromosomes. It was, therefore, essential that DNA be present in the centriole, and homologous to the centromeres.

Unfortunately, supporters of this idea persisted for a long time. Even in 1989, a paper appeared in *Cell*, one of the top journals of biology, stating that the basal body of *Chlamydomonas* contained DNA (Hall et al. 1989). After extensive, worldwide re-examination by many researchers, this "discovery" turned out to be false (Johnson and Rosenbaum 1991), but the original Cell paper was not withdrawn. The replication of centriole, which is a protein complex, remained unknown until recently. For

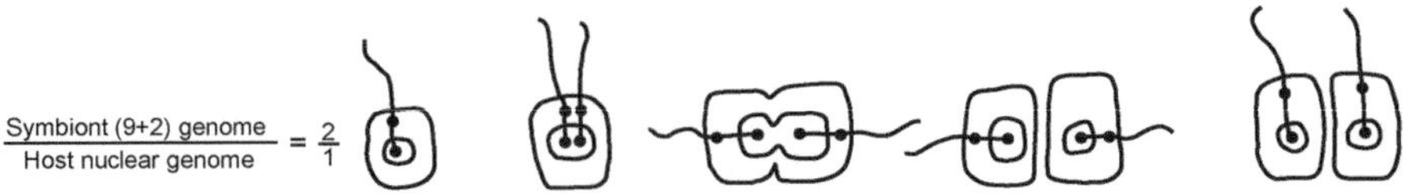

DIAGRAM V

Fig. 4.13 Relationship of flagella, centrioles, and mitosis according to the idea of Sagan (1967). Each string represents "(9+2) structure," and each black dot its hypothetical genome (DNA). Note that, after the duplication of the "(9+2) structure" in the evolution, a copy of DNA remained in the basal body, while another copy was supposed to locate in the centromere. Chromosomes are not shown explicitly, but the nucleus is shown as a circle. Only a single centromere (black dot) is shown in a cell. Mitosis proceeds from left to right. The third diagram showed metaphase, having a single set of centromeres, and the spindle was not shown in its correct form. Although Margulis did not explain the diagrams or the meaning of string and dot in the text (of neither Sagan 1967, Margulis 1970, nor Margulis 1981/1993), I deciphered the diagram in this way. Because of this strange drawing, this diagram is hard to understand for anybody (now and then). This diagram was an explanation of principle, like physics, and was not suitable in biology. In addition, the presence of endosymbiont DNA in basal body and centromere is scientifically false. The explanation of Margulis had, therefore, fatal error in its essence. The problem was not whether this was a spirochete endosymbiosis or not. Similar diagrams were shown in both Sagan (1967) and Margulis (1970). (Explanation revised from the Japanese version) Reproduced with permission of the publisher

reviews on the molecular understanding of centriole replication, see Fu et al. (2015) and Kitagawa (2017J).

The strange logical structure of Margulis' idea on the roles of (9+2) structure in mitosis seemed too difficult to understand for those who read her writings, and was not subject to criticism. Major criticism was targeted to her idea that the origin of (9+2) structure was a spirochete. However, the essence of Margulis' hypothesis was the notion that endosymbiosis of (9+2) structure was the origin of mitotic mechanism. This was an explanatory principle of the origin of eukaryotic cells, which was the title of her 1970 book. In this situation, endosymbiotic origin of mitochondria and chloroplasts was unimportant details. She described various different types of mitosis in different eukaryotic cells, which she positioned in the history of eukaryotic evolution. Importantly, she believed that these different cells acquired independently blue-green algal cells that turned into chloroplasts of different colors. On the origin of mitochondria, she thought that an aerobic bacterium was incorporated into an ancestral host cell, which became a eukaryotic cell that lives by respiration. According to Margulis, aerobic bacteria appeared by mutation of anaerobic bacteria as an adaptation to aerobic environment created by oxygen evolution of blue-green algae.

Curiously, Margulis explained the evolution of almost everything, such as photosynthesis and aerobic respiration, by accumulation of mutations, except the birth of eukaryotic cells, in which symbiosis was mandatory. She did not consider that accumulation of mutations could generate mitochondria and chloroplasts. It is difficult to understand why she had to use two different types of explanations. This question that I raise might not be evident for the readers accepting the endosymbiotic theory, but if we read the original text of Margulis in the original context and the scientific atmosphere of the time (namely, in the context of biology that I learned as a student), we will have a difficulty in understanding the need for the two opposing explanations. According to Margulis, photosynthesis was established in blue-green algae by accumulation of mutations. Why did she deny that eukaryotic photosynthesis could evolve by accumulation of mutations? In other words, the reason remained unclear why she believed that photosynthesis could not be realized in eukaryotes by accumulation of mutations. Conversely, why did she believe that blue-green algal endosymbiosis was necessary for a eukaryotic cell to acquire photosynthetic ability? Present-day researchers can easily answer the question by pointing out various similarities of photosynthesis in blue-green algae and chloroplasts. Namely, we have ample molecular data showing that both have very similar protein compositions of Photosystem I and Photosystem II. It is hard to believe that such similar mechanism of photosynthesis evolved independently, and the data cannot be explained without assuming endosymbiosis. This is the current view. In the days of Margulis, however, molecular data were not available. Production of oxygen and existence of chlorophyll were the major evidence for the similarity of blue-green algae and chloroplasts. In the biology of 1960s, we cannot find strong evidence that supported a common origin of photosynthesis in blue-green algae and chloroplasts. Even if such evidence existed, Margulis would not be able to use it, because she was not a specialist in photosynthesis. It should also be emphasized that methods of molecular phylogenetics did not exist then, which can distinguish between homologous evolution and convergent evolution in assessing similar characters.

The evolutionary thoughts in the 1960s and 1970s were still based on evolutionary synthesis originating from Darwinism. They were reinforced by increasing knowledge at the molecular level, and researchers believed that molecular mutations including base changes, insertions and deletions, could explain not only minor changes in characters but also major evolution accompanying morphological changes. Jacques Monod (1970) explained in his best-seller "Le hasard et la nécessité" (*Chance and Necessity*) that accumulation of mutations or base changes in DNA was subject to natural selection to engender well-designed proteins that function in a teleonomic manner (having a high fitness value). In the current understandings, macroscopic evolution might not be achieved by accumulation of single base mutations. Various ideas such as Evo-Devo (evolutionary developmental biology) try to explain macroscopic evolution. I explained the changes in evolutionary thinking from the 1960s in one of my work *40 years after "Chance and Necessity". Reconstruction of Monod's view on life, evolution, and the human future* (Sato 2012J). In the 1970s, molecular biology was almighty, stating that only gene mutation was a source of evolution. Margulis also explained most evolutionary events according to this hypothesis, affirming Darwinian evolution by mutation and selection in her writings. Her standpoint was, therefore, a strange amalgam: on one hand, she emphasized absolute necessity of symbiosis to explain eukaryotic origin. On the other hand, she accepted that accumulation of mutations could explain various new characters, the idea shared by many biologists of the time, reflecting the power of rapidly progressing biochemistry and molecular biology. In my opinion, the central objective of Margulis, namely, to consistently explain the whole evolutionary history of eukaryotic organisms by the evolution of mitotic mechanism, necessitated exclusion of mitochondria and chloroplasts from eukaryotic evolution. To this end, she affirmed endosymbiotic origin of these organelles. As a result, she presented a logical inconsistency of permitting two different explanations of evolution. This could have been a result of her strong belief in her evolutionary scenario of mitosis.

4.3.2 Multiple Parallel Endosymbiosis

In her 1970 book, Margulis presented a contrast between the fossils before and after the Cambrian era, as well as fundamental distinction between eukaryotes and prokaryotes. According to the knowledge of paleontology of the time, no large organisms existed in the pre-Cambrian time. Only microfossils were known. Margulis considered that transition from prokaryotes to eukaryotes occurred around the Cambrian era. She did not argue about the differences between unicellular and multicellular eukaryotes. She did not give detailed discussion on the differences between plants and animals. She seemed to consider simply that plants were animals that captured chloroplasts. Accordingly, both Mereschkowsly and Margulis gave a similar explanation in which diversified aerobic eukaryotes acquired differently colored blue-green algae to give various plants and algae that we find now on the Earth. The figure shown as a summary in Mereschkowsly (1910) was essentially similar to the frontispiece figure in Margulis (1970). Figure 4.14 is intended to show essential idea of Margulis for comparison with Mereschkowsky's image. Margulis compared Mereschkowsky's scheme with hers later (Margulis and Bermudes 1985).

We discussed "botanical myth" in the first section of this chapter. This was a result of the fact that algal evolution was traditionally analyzed by focusing on plastids. If we take cyanobacteria and plastids as a measure of evolution, an evolutionary path from cyanobacteria through red algae and green algae to plants might not be wrong. Since photosynthetic properties were the best studied in botany in the 1960s, we can understand why such path of evolution was taken as a default. In contrast, Margulis proposed a different path of evolution in which evolution of eukaryotes proceeded first without chloroplasts, which were acquired after the diversification of eukaryotes. I invite the readers to examine carefully the number of endosymbiotic events in the very large figure of phylogenetic tree (Figure 2–7) in *Origin of Eukaryotic Cells*. A simplified version is shown in Fig. 4.15. The circles indicate acquisition of plastids, with a note indicating the color of acquired plastid.

Marugulis did not provide explanation of the presence of various, diversified blue-green algae to be acquired by various eukaryotic cells. She did not evaluate the possibility of 15–20 independent endosymbioses of such blue-green algae. Although Margulis repeated that endosymbiosis was the most rational explanation with a

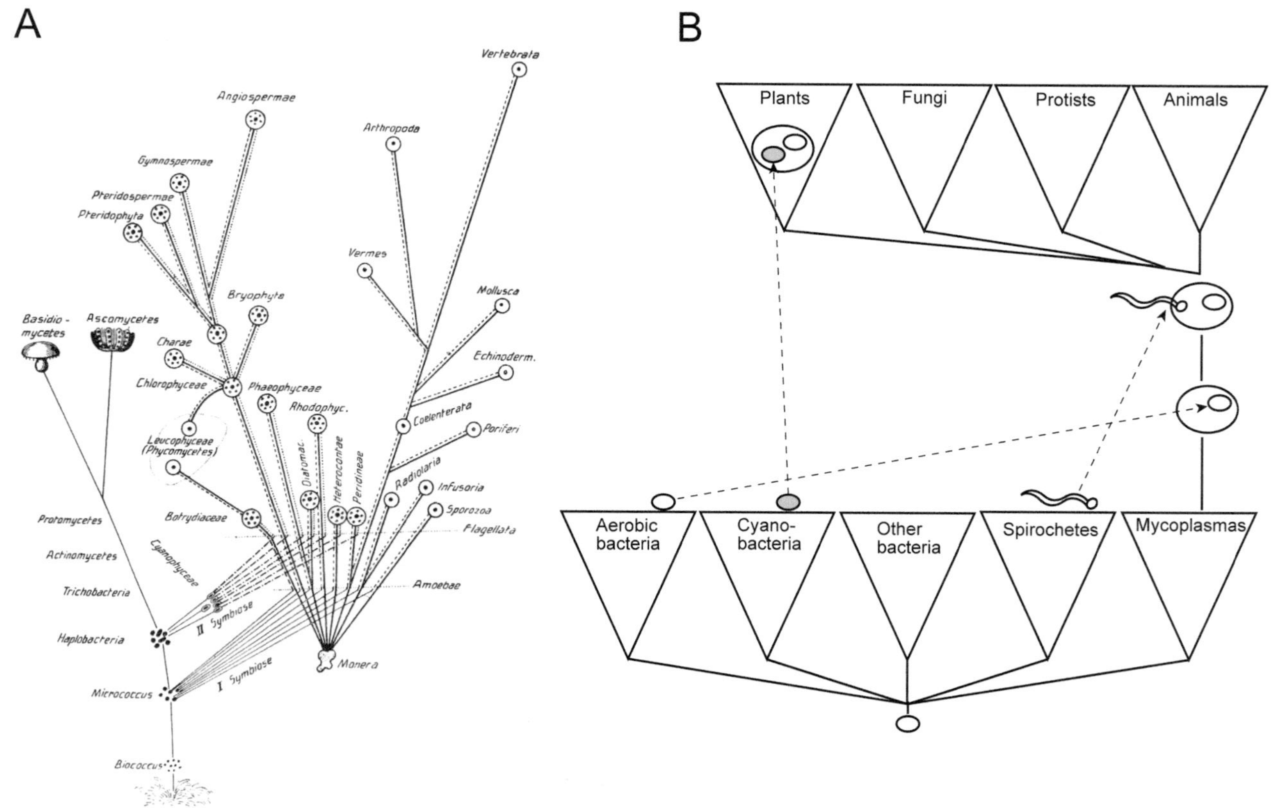

Fig. 4.14 Comparison of the diagram of Mereschkowsky (**a**) and a simplified diagram of Margulis (**b**) showing the endosymbiotic evolution. Mereschkowsky did not describe mitochondria. He described that the initial endosymbiosis resulted in cell nucleus, and the second plastids. The diagram in (**b**) is a simplified figure explaining the essential idea of Margulis, which was redrawn after the frontispiece of Margulis (1970). Margulis did not show multiple endosymbioses of plastids there. The design of the two diagrams was curiously similar

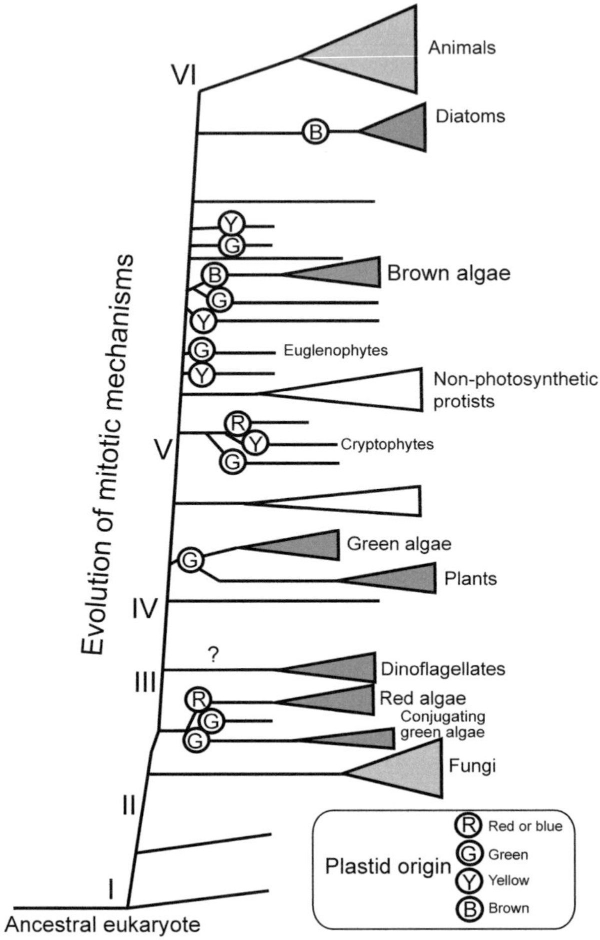

Fig. 4.15 Diversification of eukaryotic cells and acquisition of plastids in different lineages according to Margulis. This is a simplified figure explaining the idea of Margulis (1970). The roman numbers I–VI indicate the beginning of the six steps in evolution of mitosis. As noted in the bottom right box, circles indicated acquisition of plastids. Namely, four types of blue-green algae, R, G, Y, and B, were incorporated into various eukaryotes to become different plastids of different colors. In this figure, 15 independent endosymbiotic events are drawn. Evidently, yellow and brown "blue-green algae" do not exist. This is quite different from the currently accepted phylogenetic tree (Sect. 6.3): animals are placed on top, while fungi are placed at the bottom. Although Margulis explained that her tree was drawn according to Copeland (1956) and Whittacker (1969), Margulis seemed to modify it (NB: these two citations are not included in the reference list of this book). The branching in this figure is also quite different from the phylogenetic trees that Margulis presented elsewhere. In the text of Margulis (1970), different mitotic mechanisms of plants and green algae were described, but these two were placed in the same lineage resulting from a single event of green plastid acquisition. Margulis did not show any endosymbiotic event in the origin of dinoflagellates. Phylogenetic trees were drawn quite arbitrarily before the introduction of molecular phylogeny. Original figure after Margulis (1970)

minimal number of evolutionary steps, multiple parallel endosymbiosis assumed many more steps. Phylogenetic relationship based on plastids demonstrated rather validity of "botanical myth." We might be able to consider that plastids wear a mantle of eukaryotic cell rather than a eukaryotic cell ingests a plastid. Looking at the cells of unicellular red alga, *Cyanidioschyzon merolae*, in which a single plastid occupies more than a half of the cell volume, we are tempted to imagine a plastid domesticating a eukaryotic nucleus. In contrast, the explanation by Margulis assuming that diversified eukaryotic cells acquired diversified blue-green algae has a degree of freedom too high to obtain supporting evidence. Indifferent to the number of endosymbiosis events, Margulis kept aside plastid endosymbiosis by focusing on the main topic, mitosis.

4.4 Curious Citations of Mereschkowsky in Margulis's Publications

In her 1967 paper and 1970 book, Margulis did cite some papers in the beginning of the twentieth century including Mereschkowsky, but the citations remained indirect. Mereschkowsly's papers were not cited correctly even in the late 1970s or later when Margulis became established, leading scientist (Table 4.4). I already pointed out this problem in a recent paper (Sato 2017). Although some researchers might have noticed this problem, no one has ever openly discussed it, maybe because this could be an act of impoliteness. The only correct citation of Mereschkowsky that I know was the citation of the 1910 paper in Margulis and Bermudes (1985). It is interesting to note that this 1985 paper presented the figure of Mereschkowsky (1910) and compared it with the evolutionary scheme of Margulis. This could have necessitated correct citation of Mereschkowsky's paper.

I wonder if incorrect citations of Mereschkowsky could have contributed to the general recognition of her achievements. I do not know whether she did it intentionally or not. In any way, this had no significant effect in the discussions until the mid-1970s, because all researchers were aware of the historical arguments of endosymbiosis including Mereschkowsky and Famintzin, and it was not necessary to cite Margulis. Indeed, her works were rarely cited in the papers on endosymbiotic discourses of the time.

4.5 Integration of Most Recent Findings in Other Disciplines

4.5.1 *Cell Biology/Biochemistry*

Many researchers affirmed that Margulis gathered modern data to establish a modern version of the endosymbiotic theory, which had been a fantasy until then. One of the reasons for this was certainly their belief in Margulis' statement in her 1970 book:

Table 4.4 Citations of Mereschkowsky by Margulis (modified from Sato 2017)

ID	Publication of Margulis and description of citation
1	**Sagan** (1967) MERESCHKOWSKY, M. (1910) & MINCHIN, E. A. (1915). *In* "The Cell in Development and Heredity", by E. B. Wilson (1925). New York: Macmillan. [This is a secondary citation through the book of Wilson, and the spelling of Mereschkowsky, although there were various spellings of Russian name, was incorrect]
2	**Margulis** (1970) [Mereschkowsly was not in the reference list]
3	**Margulis** (1975) [Mereschkowsly was not in the reference list]
4	**Margulis and Bermudes** (1985) Mercschkowsky, K. 1905. Le plante considéré comme une complex symbiotique. *Bull. Soc. Nat. Sci., Quest* 6: 17–98. [Incorrect citation of Mérejkovsky (1920). "*Quest*" should be "*Ouest*": Many errors in the French title. Correct citation should be: 1920. La plante considérée comme un complexe symbiotique] Mereschkowsky, K. 1910. Theorie der zwei Plasmaarten als Grundlage der Symbiogenesis, einer neuen Lehre von der Entstehung der Organismen. *Biologische Centralblatt* 30: 352–367. [This is correct]
5	**Margulis** (1981/1993) Mereschkovsky, K. C. 1905. Le plant Considéré comme un complex Symbiotique. *Bull. Soc. Nat. Sci., Ouest* 6:17–98. [Similar errors in (4)] Mereschkovsky, K. C. 1909. Theory of two plasms as the basis of symbiogenesis, new studies about the origin of organisms. USSR: Kazan. (in Russian, cited in Takhtajan 1973). [Curiously, only a Russian paper was secondarily cited through another Russian article for a popular scientific magazine. Mereschkowsky was mentioned very briefly in the text of Takhtajan (1973). Note that Mereschkowsky published the theory of two plasmas in German in the next year as listed in (4)]
6	**Margulis** (1998) [Mereschkowsky was mentioned in the text many times, but none of the works of Mereschkowsky was cited in the reference list containing 37 references]

"The symbiotic theory developed in this book challenges this tacit assumption (=that the evolutionary mechanisms deriving eukaryotes from prokaryotes are the same as those operating within each of these great groups of organisms), using *modern biochemical, genetic, cytological, and paleontological information*" (Margulis 1970, p. 5. My emphasis). "Modern biochemical information" included analytical data of DNA, protein synthesis systems, various metabolic enzymes, and respiratory/photosynthetic systems. "Genetic information" included well-known cytoplasmic inheritance of plastids and mitochondria, which were based on the review article of Lederberg (1952). "Cytological information" included electron microscopic observation of organellar DNA and organellar ribosomes. "Paleontological information" included drastic changes in fossils before and after the Cambrian era, as well as the rise in oxygen partial pressure about 2.7 Gya during the Earth history.

Unfortunately, the usage of these data was not adequate from the current viewpoint: first, fossil data were used to locate prokaryotic/eukaryotic transition in the Cambrian era. Second, emergence of aerobic bacteria was assumed after the rise in oxygen concentration. These points are now incorrect. Although Margulis listed biochemical, genetic, cytological, and paleontological information, the most important evidence was the presence of organellar DNA. The biochemical information that she described, including metabolic pathways and various compounds such as chlorophyll, carotenoids, tocopherols, sterols, unsaturated fatty acids, and cytochromes, was used to explain emergence of aerobic respiration and survival of oxidative stresses, which are no longer valid.

We can trace how newly acquired data were utilized to substantiate endosymbiotic theory in the papers that also proposed endosymbiotic theory of chloroplasts and mitochondria in the 1960s and 1970s. These papers provided detailed tables that compared characteristics of organelles and bacteria (see Chap. 5), which were cited and reproduced in Margulis (1970). Margulis herself provided only a few scientific data that could support endosymbiotic theory of chloroplasts. Margulis must have understood at the time that she did not have much scientific contribution in proposing endosymbiotic theory of chloroplasts and mitochondria.

Curiously, although Margulis did not have scientific works on microtubules and mitosis, she was able to assemble many data to construct an original hypothesis involving endosymbiotic origin of (9+2) structures, which was the core of her thoughts. At the time, a considerable number of papers were published on the presence of DNA in centrioles. Many papers provided electron micrographs of centrioles, flagella, and spindle microtubules. These provided "information" available in the cytological discipline in which Margulis was educated. In addition, inhibitors of cell division were known. When she wrote the phrase "challenges the tacit assumption using cytological information," she was certainly arguing about endosymbiotic acquisition and evolution of (9+2) structures, rather than the endosymbiotic origins of chloroplasts and mitochondria, which were already formulated by other researchers.

4.5.2 Earth Science

In the 1960s, not only molecular and cell biology, but also earth science showed a significant progress. New understandings on the history of life on the Earth were provided by Bernal (1967) (see Figure 2 therein: "Diagrammatic visualization of evolution of the simple living cell"). We are surprised to see how rapidly Sagan (1967) and Margulis (1970) incorporated new findings of Earth science (see Fig. 4.16).

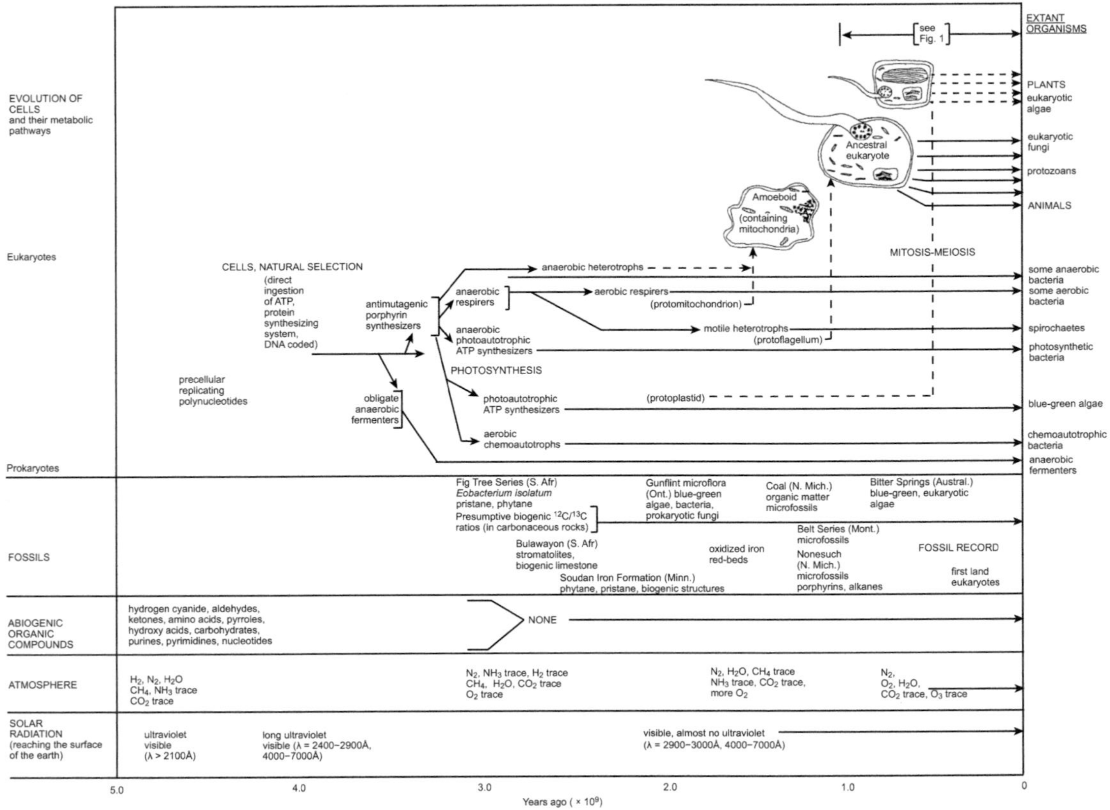

Fig. 4.16 The history of life according to Sagan (1967). A similar, but renewed figure is found in Margulis (1970). Reproduced with permission of the publisher

These included the following findings: Life emerged on the Earth about 3.5 Gya (billion years ago); oxygen concentration was very low in the early times but increased rapidly from about 2.7 Gya; emergence of respiration after photosynthesis; no significant eukaryotic fossils were found before Cambrian explosion of biodiversity (500 Mya). All these were novel knowledge of the time that Margulis described in Sagan (1967) and *Origin of Eukaryotic Cells*. It was certainly difficult for top molecular biologists to keep up-to-date knowledge on the progress in a different discipline. J. D. Bernal was a rare biochemist having knowledge in molecular and structural biology, who also had a strong interest in the origin of life. He was able to use newest data of Earth science to construct a realistic view on the origin and evolution of early life. As a pioneer in the studies on the origin of life, Bernal reproduced the publications of Oparin and Haldane as supplements of his 1967 book. Bernal already knew Margulis' 1967 paper before its publication, and cited it as a 1966 publication in the reference list (Bernal 1967). However, he cited it in the context of origin of eukaryotic cells, namely, that prokaryotes and eukaryotes are fundamentally different. Bernal described cell organelles and their possible origins according to autogenous or endosymbiotic hypotheses, but he remained prudent regarding the endosymbiotic theory of organelles:

> The evidence for this idea [=endosymbiotic hypothesis] is still all of an inferential character. No independent organelle has been found existing as a free organism, but the structure of all cells down to the bacteria is found to contain organelles. Further, these organelles seem to have an existence of their own and often to be largely independent of the central genetic control of the reproductive organism. They seem to behave as if in some way they are a kind of foreign element introduced into the organism, and several in eukaryotes have marked resemblances to free-living prokaryotic cells. Blue-green algae are sometimes almost completely indistinguishable from 'plastids.' (Bernal 1967, p. 81)

As Margulis stated, Bernal suggested that photosynthetic oxygen production lead to evolution of aerobic respiration:

> This transformation from prokaryotic to eukaryotic cells, which may have occurred rather late in the evolution of life, may, indeed, not have occurred all at one time but spread over a number of millennia or aeons. One decisive change was the occurrence of photosynthesis, liberating free oxygen and requiring appropriate biochemical changes of which the replacing of fermentation by oxidation is the principal one. (Bernal 1967, p. 87)

Bernal described the origin of plants in terms of endosymbiosis:

> The eukaryotic plant cells seem definitely to be of multiple origin, engulfing and maintaining photosynthetic plastids, originally free prokaryotic organisms. (Bernal 1967, p. 88)

After reading clear and insightful explanations of Bernal, we have a simple question what Margulis added to Bernal's explanation. It is quite probable that Bernal, a leading biochemist of the time, provided a prototypic hypothesis on the endosymbiotic origin of plastids as a whole. In addition, Bernal himself was not an enthusiastic supporter of the endosymbiotic theory as shown above. In this situation of general hesitation, Margulis could have played the role of pushing the people to believe the endosymbiotic theory by her strong confidence.

How did young Lynn at the age of 29 obtain newest knowledge of Earth science? It was certainly Carl Sagan, the first husband of Lynn, who provided it. According to her later reminiscence (Margulis 1998), a student Lynn Alexander fell in love with Carl, a smart and intelligent, elder student. Her 1967 paper cited papers of Carl, suggesting that the couple was discussing the origin of life at that time. Unfortunately, after their divorce, no paper of Carl Sagan was cited in Lynn's publications, but she must have acquired invaluable intellectual gift from him. Lynn Margulis represented herself as a young, talented biologist, not just familiar with biochemistry, morphology, and taxonomy, but endowed with an exceptional ability of manipulating knowledge in various different fields of science.

Chapter 5
Endosymbiotic Discourses in the 1960s and 1970s

As described in Sect. 4.5, the progress of sciences in the 1960s and 1970s had significant influences on the advances in the endosymbiotic hypothesis. Above all, discovery of organellar DNA, visualization of intracellular structure by electron microscopy, and elucidation of mitotic mechanism, all these were important progresses that formed the basis of modern cell biology. We will overview the history of endosymbiotic hypotheses in these decades, by referring to Sapp (1994) as necessary.

5.1 General Description of the Trends of the Time

In 1963, during the Cuban Missile Crisis, the Society for General Microbiology held a symposium on symbiosis in London. The proceedings were published as *Symbiotic Association* (Nutman and Moose 1963). In the Preface, the editors described serious dilemma of symbiotic studies:

> It has often been maintained that the subject of symbiosis has no independent standing, and that its name tends to disguise a fundamentally parasitic relationship – surrounding its admitted complexity with an unscientific aura of teleology. (Nutman and Moose 1963, p. ix)

According to Sapp (1994, p. 167), "the emphasis on the negative effects of symbiotic associations was, of course, by no means simply a reflection of Western ideology. It was reinforced by medical constructions of the inherent pathological nature of bacteria and viruses." Nevertheless, Dubos and Kessler (In Nutman and Moose 1963, p. 1) emphasized "creative manifestations of symbiosis." Sapp (1994, p. 167) paraphrased it as "creative dynamics of symbiotic relationships." Dubos and Kessler directed the discussion to biological context (ecological advantage) rather than pathological one. They described curious properties of the petite mutants of yeast affected in mitochondria and bleaching (loss of chloroplasts) of *Euglena* cells treated with streptomycin as interesting examples of drastic changes in characters

N. Sato, *Endosymbiotic Theories of Organelles Revisited*,
https://doi.org/10.1007/978-981-15-1161-5_5

due to organelle losses, and inferred the origin of eukaryotic cells as a symbiotic assembly. Except lysogenization of bacteriophage, topics of the 1963 meeting remained largely limited to classical examples, such as lichens, root nodules, and mycorrhiza (fungal association of plant roots).

Nevertheless, Sapp (1994) recognized the value of green paramecia investigated by Karakashian at the time as a model of endosymbiosis. In this context, the genetic approach to symbiosis had three virtues (Karakashian and Siegel 1965): synthetic power, genetic and biochemical integration, and heuristic value in research. By contrast, Buchner (1953/1965) denied symbiotic origin of eukaryotic cells (Sect. 3.7). Sapp (1994, p. 170) commented on the atmosphere of the time: "When referring to the speculative nature of such theories Buchner expressed a common sentiment among biologists who believed that such origin stories were simply not scientific, in the sense that one could not prove or disprove them." As described above (Sect. 3.7), Buchner's attitude was more negative, and nearly hostile to the endosymbiotic origin of organelles.

Sapp pointed out that the "space race" during the Cold War was a reason that promoted scientific origin search. The spaceship Sputnik launched by the Soviet Union had an official aim of exploring the origin of space and life. Space biology was also a theme of space exploration of NASA. These fostered an atmosphere tolerant to hypotheses on the evolution of life. In my opinion, space exploration was fundamentally military in nature, but this essential aspect was hidden behind a pacific, cultural target, namely, origin of life. We must recall that the research on the origin of life remained at a very primitive, nearly "pseudoscientific" stage at the time, but was used or abused for space exploration. Recent space biology seems much more scientific. In contrast, the space biology in the 1960s was nothing more than a camouflage of arms race in the Cold War.

5.2 From the Biological Revolution to the Endosymbiotic Theory

Development of molecular biology starting with the double helix model of DNA proposed in 1953 became a major, big stream in the 1970s, which promoted progress in entire biological researches. This can be called biological revolution in contrast with physics revolution in the early twentieth century. The progress in biology as a whole was necessary to recognize and accept the endosymbiotic hypothesis of organellar origin, which was not just a simple story of incorporation of cyanobacterial and proteobacterial cells within a eukaryotic host, as originally imagined in the early twentieth century.

5.2.1 Establishment of Cellular Ultrastructure

Until the early 1960s, only optical microscope was available for biological observation. Details of intracellular structure and existence of plasma membrane remained unclear. Electron microscope, which had been developed before the Second World War, became applied to biological specimens after the war to elucidate cellular structure. This allowed microbiologists to visualize internal structure of unicellular microbes and bacteria, which had been considered primitive. It also enabled comparison of organelles and bacteria at high magnification.

Distinction between eukaryotes and prokaryotes was proposed by a French biologist Édouard Chatton in 1937, but limited resolution of optical microscope remained an obstacle to clearly demonstrate the distinction. Stanier and van Niel (1962) established cellular organization of bacteria, stating "*all bacteria are cellular entities*" (p. 19. Italic in original text). They authorized the use of "cell" to both eukaryotes and prokaryotes. In addition, they re-established the distinction between eukaryotes and prokaryotes, based on the internal cellular structure. Nucleus, mitochondria, and chloroplasts were major characteristics of eukaryotic cells. Prokaryotic cells had a DNA-containing region, but did not possess nuclear membranes. Thus, cyanobacteria (called blue-green algae then) and bacteria were prokaryotes, and mitosis was specific to eukaryotes. Stanier and van Niel (1962) considered that mitosis was not necessary in prokaryotes, which had a single chromosome. In 1961, fine structure of chloroplasts, consisting of envelope membranes, lamellae (later called thylakoids), and stroma, was elucidated.

Interestingly, the article described mitochondria as "respiratory plastids," which indicated persistent ambiguity in plastids and mitochondria. Absence of mitochondria in aerobic bacteria was confirmed. Mesosomes, invagination of inner membrane, were recognized in bacteria. A blue-green algal cell included lamellae without envelope membranes, thus homologous to a chloroplast. Functional localization of respiration and photosynthesis to mitochondria and chloroplasts, respectively, was especially emphasized, as an achievement of biochemical studies using isolated organelles beginning from the 1950s. This was a good indication of functional differentiation within eukaryotic cells. The basic biological knowledge like these, which is now described in introductory biology textbooks, was first recognized in the 1960s, which enabled comparison of mitochondria and bacteria or that of chloroplasts and blue-green algae. *The Microbial World* by Stanier et al. (1957/1976) was thus an epoch-making textbook that formed the foundation of modern microbiology.

A long review article (more than 40 pages) of Echlin and Morris (1965) described detailed ultrastructure of blue-green algae, and concluded: "bacteria and blue-green algae should be isolated in a group distinct from all other organisms" (p. 177). Young readers should recognize that very basic, common knowledge of biology like this was being established in the 1960s. In this sense, the standpoint in which prokaryotes were clearly distinguished from eukaryotes, especially the view in which blue-green algae should be grouped with bacteria rather than plants, was quite novel at the time. Under these circumstances, the critique of "botanical myth" by Margulis, namely, refutation of phylogenetic link between blue-green algae and eukaryotic algae, can be considered a hasty and premature action for a young novice in the research field. Discussions still continued at the time.

5.2.2 *Discovery of Organellar DNA*

There was a long history of research on cytoplasmic inheritance. Not many people were working in genetics of organelles in the 1960s. Ralph P. Lewin (1921–2008) and Ruth Sager (1918–1997) were pioneers in a new research field, chloroplast genetics, using a green alga *Chlamydomonas reinhardtii*. This microalga, currently extensively studied as a model green alga to develop biofuels, was first described by P. A. Dangeard (1862–1947) in 1888, but the ancestor of laboratory strains was isolated from a potato field by G. M. Smith (1885–1959) in 1945, and then maintained as three different lines in different laboratories (namely, Sager line, Cambridge line, and Ebersold/Levine 137c line: see *The Chlamydomonas Sourcebook* [Harris 2008] for history and current research topics). Chloroplast DNA of *Chlamydomonas* was first detected by electron microscopy by Ris and Plaut (1962), and then isolated by ultracentrifugation by Sager and Ishida (1963). Ris and Plaut (1962) suggested endosymbiotic origin of chloroplast by citing Mereschkowsky (1905a), whereas Sager and Ishida (1963) wrote nothing about it. Sager never referred to endosymbiotic hypothesis in her later works (see, for example, Sager 1967). Margulis criticized Sager's works many times (see, for example, Margulis 1970).

Mitochondrial DNA was detected by electron microscopy by Nass and Nass (1963a, b), who referred to endosymbiotic origin of mitochondria. Nass et al. (1965a) established isolation of mitochondrial DNA. Meanwhile, chloroplast DNA was detected or isolated in various algae and plants. GC content of chloroplast DNA was estimated, based on buoyant density measured by ultracentrifugation over CsCl density gradient (Edelman et al. 1965, 1967). These authors compared buoyant density of DNA in blue-green algae with that of nuclear and chloroplast genomes of algae and plants. They also provided data on the nuclear and chloroplast DNA in a glaucophyte *Cyanophora paradoxa*, which has been known to contain cyanelles having peptidoglycan (see Chap. 6, Fig. 6.26). Significant difference in buoyant density of nuclear and chloroplast DNA was taken as evidence against autogenous origin of chloroplasts. However, large differences in buoyant density of DNA were detected in different groups of blue-green algae. Therefore, different blue-green algae were assigned as origins of different chloroplasts depending on different groups of plants and algae. I have to remind the readers that GC content is not a good measure of estimating phylogenetic relationship, and that the discourses described above on the GC content of blue-green algae and plastids was useless in establishing endosymbiotic origin of organelles.

5.2.3 Discourses on the Endosymbiotic Origin of Organelles

Discourses continued on the endosymbiotic origin of organelles. Klein and Cronquist (1967) wrote a long review on the phylogeny and evolution of bacteria, algae, and fungi. They presented data on the Earth history, ideas on the origin of life, biochemical comparison including porphyrins, cytochromes, chlorophylls, phycobilins, carotenoids, carbohydrates, lipids, and other compounds, cytological characteristics (including the 9+2 structures!), and photosynthetic properties, just as Margulis presented in her 1967 paper and 1970 book. But they were quite prudent in assessing evolutionary history, and criticized Ris for circulating "bad penny":

> Ris (1962) has demonstrated the presence of DNA in the chloroplasts of several algae. He concluded that this DNA represents the "genetic system of the chloroplast" and suggested that the plastid of *Chlamydomonas* (and by extension, all other algae, and possibly higher plants) may be an endosymbiont of an original blue-green alga. This bad penny has been circulating for a long time. Ris, and others before him who have made the same suggestion, have apparently ignored the fact that chlorophylls in addition to chlorophyll *a* are characteristic of the photosynthetic apparatus of most other algae and higher plants, that bilin pigments are rarely found in the plastids of plants other than blue-green (and red) algae, and that the structure of the plastid of plants beyond the blue-green algae is fundamentally different in possessing a plastid membrane which is not at all like the cell membrane of blue-green algae in composition and structure. Clearly, there is no chemical, structural, or phylogenetic basis for this belief. (Klein and Cronquist 1967, p. 167)

The endosymbiotic hypothesis of organelle origin was discussed in Symposium of the International Society for Cell Biology in 1966, and the proceedings were published next year (Warren 1967). Some authors did question the hypothesis. Among the participants, Gibor (1967) summarized the arguments for and against the hypothesis. Arguments that supported endosymbiotic origin of organelles:

> (1) Different degrees of endosymbiotic relationships are known; (2) the DNA of the plastids and mitochondria differs qualitatively from the nuclear DNA; (3) the plastids and mitochondria are separated by a double membrane from the rest of the cell; (4) the organization of the DNA fibers of the organelles, as revealed by electron microscope studies, resembles the DNA organization in a bacterial nucleoid rather than the chromosomes of the nucleus of higher organisms. (pp. 312–3)

There were a number of problems, too:

> (1) the fact that the organelles of different species differ in their DNA would require that different symbionts enter different cells and all proceed to evolve in an identical pattern; (2) some of the enzymes of the organelles are coded for by nuclear genes; (3) some of the RNA of the mitochondria appears to complement nuclear DNA, suggesting that it was synthesized in the nucleus and transferred into the mitochondria. (p. 313)

He indicated "a possible advantage to the cell in the proliferation of only parts of its genome," citing examples of bacterial episome-like DNA, and rDNA of higher organisms. He considered these "an efficient and rapid mechanism for the synthesis of essential proteins" and evolutionarily advantageous:

> Ontogenetically the cell is provided with an efficient and rapid mechanism for the synthesis of essential proteins; phylogenetically the multiplicity of templates can function to stabilize very important enzyme systems from the hazards of deleterious mutations. (p. 314)

According to Sapp (1994, p. 174), "this good fit of form and present function permitted in itself no conclusion about how these organelles or any other organismic structure originated." He cited the famous satire of Leibnitzian belief of Dr. Pangloss in Voltaire's *Candide ou l'Optimisme* (p. 2), "tout est nécessairement pour la meilleure fin." (everything is made for the best purpose). Note that this is a frequently used metaphor in philosophy of science to criticize the belief that various traits of organisms can be explained by optimizing action of natural selection (see, for example, Morange 2016, p. 352).

Haldar et al. (1966) demonstrated that cytochrome *c* is synthesized in the cytoplasm. Based on this finding that invaginations of bacterial membrane called mesosomes bound DNA, they proposed a hypothesis that a part of chromosomal DNA was transferred to mesosomes to build centrioles, mitochondria, kinetoplasts, basal bodies, and plastids. This was a prototype of autogenous hypothesis. Supposedly, this hypothesis was difficult to construct without detailed knowledge of internal cellular structure. Autogenous hypothesis must have been invented hastily and forcibly by those who were reluctant to accept endosymbiosis hypotheses. The difficulty in theoretical formulation of autogenous hypothesis was reflected in its various different forms depending on researchers.

5.3 Researchers That Proposed Endosymbiotic Theory of Organelle Origin

In this section, I will describe (or re-discover) various researchers that proposed endosymbiotic origin of organelles. A microscopist in Cambridge, Patrick Echlin (1932–2015) presented a detailed comparison of blue-green algae and chloroplasts (Echlin 1966), and suggested a possibility of endosymbiotic origin of chloroplasts (Table 5.5). We are impressed with detailed comparison of the two entities. He wrote a review article on the ultrastructure of blue-green algae (Echlin and Morris 1965), which made this table possible. A corresponding table of comparison between mitochondria and bacteria was constructed by Nass et al. (1965b). Echlin wrote several additional reviews. In Chap. 1 of *Phytochemical Phylogeny* (Echlin 1970), he described endosymbiotic theory of chloroplasts by citing his own article but not Sagan (1967). In contrast, in Chap. 7 of the same book, Carr and Craig (1970) presented various data related to endosymbiotic hypothesis by citing Sagan (1967) among other articles. It seems that every author took his/her own position at that time.

In the next year, a Norwegian microbiologist Goksøyr (1967) proposed, in *Nature*, a hypothesis that a eukaryotic cell was formed by sequential integration of prokaryotic cells, rather than that a prokaryotic cell became a eukaryotic cell (Fig. 5.17). This was similar to the serial endosymbiosis theory (SET), which was

Table 5.5 Comparison of blue-green algae and chloroplasts (Echlin 1966)

	Unicellular blue-green algae	Higher plant chloroplasts
Size	About 6 μm	4–8 μm
Cell wall	Present. Contains diamino-pimelic acid	Absent
Osmotically active limiting membrane	Present. About 70 A across	Present. About 70 A across. Usually double
Ribosomes	Present	Present. Different from cytoplasmic ribosomes
Photosynthetic apparatus	Multilamellate. Peripherally arranged closed sacs. Each closed sac bounded by a 70–80 A membrane	Multilamellate. Closed sacs are organized into stacks of grana lamellae inter-connected by means of stroma lamellae. Each closed sac bounded by a 70–80 A membrane
Photosynthetic mechanism	Photolysis of water with consequent evolution of oxygen. Primary mechanism coupled with photosynthetic phosphorylation	Photolysis of water with consequent evolution of oxygen. Primary mechanism coupled with photosynthetic phosphorylation
Primary photosynthetic pigment	Chlorophyll *a* + accessory pigments	Chlorophyll *a* + accessory pigments
Method of reproduction	Binary fission	Binary fission? Origin de novo from preexisting membranes within the cell. Origin de novo from preexisting proplastids within the cell
Genetic apparatus	Little known. Evidence suggests it is a non-Mendelian inheritance mechanism as seen in bacteria	Evidence suggests that it consists of a non-chromosomal heredity system with an absence of Mendelian segregation ratios and of linkage
DNA	Present as 25 A fibrils, not organized into chromosomes. DNA not associated with histones	Present as 25 A fibrils, not organized into chromosomes. DNA not associated with histones. In the few cases which have been investigated chloroplast DNA has different base ratios from nuclear DNA
RNA	Present	Present. A messenger RNA complementary to chloroplast DNA has been found. A DNA-dependent RNA synthesis has been reported
Protein synthesis	Present	Present
Degree of autonomy	Entirely free living	Intact chloroplasts prepared by improved isolation techniques appear to be able to perform at least all the photosynthetic functions normally ascribed to this organelle within the cell. No data available concerning other metabolic pathways

(continued)

Table 5.5 (continued)

	Unicellular blue-green algae	Higher plant chloroplasts
Primary photosynthetic product	Present as Cyanophycean starch which appears to be equivalent to the amylopectin of higher plant starches	Present as starch, but quickly translocated to other parts of the plant

The "A" in "70 A," "25 A," and "70–80 A" means Angstrom (unit)
Reproduced with permission of the publisher

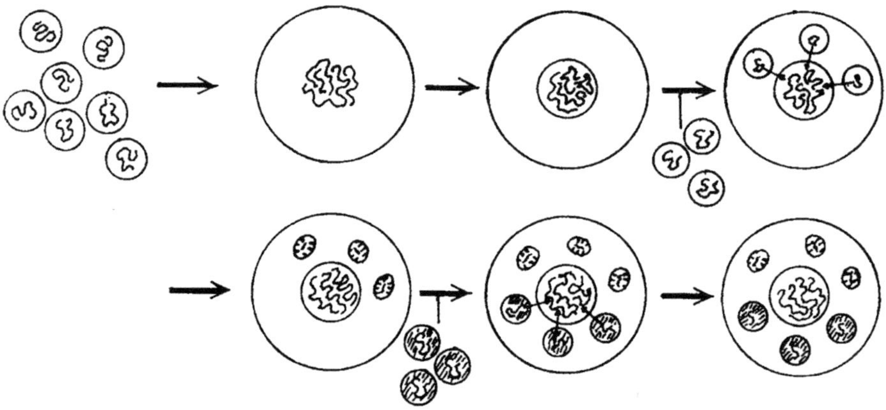

Fig. 5.17 The endosymbiotic hypothesis of organelle origin by Goksøyr. Goksøyr (1967) presented a scheme, exactly identical to the one currently regarded as established (Fig. 1.9). The first endosymbiosis generated mitochondria, whereas the second endosymbiosis produced plastids. Endosymbiotic gene transfers are also shown by arrows. Reproduced with permission of the publisher

proposed later (Taylor 1974b). In this paper, Jostein Goksøyr (1922–2000) suggested that DNA/RNA hybridization would "give some information" to prove his hypothesis. This is identical to the method of demonstration, which Margulis stated later in *Origin of Eukaryotic Cells.*

Was it possible that Echlin (1966) and Goksøyr (1967) knew the activities of Lynn Margulis (Sagan at the time) and obtained her idea? Perhaps some people that believe the priority of Margulis could doubt if the original idea of Margulis was stolen by other researchers during the processes of long, repeated reviewing of Margulis' paper to be published in March 1967. The manuscript of the 1967 paper was initially submitted in early 1966, but rejected. According to Margulis, rejection and re-submission were repeated 15 times. During the processes, researchers that were involved in reviewing the manuscript might think that it would be better to summarize arguments and data on endosymbiotic origin of organelles. This is just a possibility that might be conceived by pro-Margulis people. However, the symposium in 1966 must be planned before the initial submission of Margulis. Probably, an atmosphere of discussing endosymbiotic hypothesis was present already in 1965–1966. In other words,

Margulis did not participate in the discourses on the origins of chloroplasts and mitochondria of the time, which continued before her and without her. In this sense, addition of flagellar endosymbiosis to complete the narrative on the origin of eukaryotic cells was a good, successful strategy to emphasize her originality.

5.4 Debates Between the Endosymbiotic and Autogenous Hypotheses

In many articles published from 1965 to the early 1970s, both endosymbiotic theory and autogenous hypothesis were presented in parallel, leaving final conclusion unanswered. For example, *The Plastids* by Kirk and Tilney-Basset (1967) explained endosymbiotic hypothesis in some detail, but kept a prudent attitude without deciding which hypothesis was correct. Loening (1968) demonstrated that the size of ribosomal RNA (rRNA) was clearly different between eukaryotes and prokaryotes, and that chloroplasts contained prokaryotic rRNA. This was additional supporting evidence for the endosymbiotic hypothesis.

Stanier (1970) wrote a review article based on his lecture in the Society for General Microbiology. He compared the two hypotheses (endosymbiotic and autogenous), and confirmed that the same mechanism of photosynthesis was conserved in both blue-green algae and plants/eukaryotic algae, but left the question unanswered:

> It might have happened thus; but we shall surely never know with certainty. Evolutionary speculation constitutes a kind of metascience, which has the same intellectual fascination for some biologists that metaphysical speculation possessed for some mediaeval scholastics. (pp. 30–31)

He closed the article with an Italian phrase: "*Se non è vero, è ben trovato.*" (Even if it is not true, it sounds a good story to believe). Nichols (1970), who wrote a review chapter explaining biological substances from the viewpoint of evolution/phylogeny, described similarity of lipids between chloroplasts and blue-green algae, essentially based on the continuity of blue-green algae and eukaryotic algae. Interestingly, Carr and Craig (1970) wrote another chapter of the same book, presenting data from the standpoint of the endosymbiotic hypothesis as noted in the previous section. Baxter (1971) described possible endosymbiotic origin of organelles, but reserved conclusion, because only a small part of proteins were synthesized within the organelles. He thought that most part of proteins might have been lost if organelles were endosymbionts. Stubbe (1971) presented two possibilities, and concluded that the problem was not resolved. He cited Mereschkowsky and Schimper, but not Margulis. The presence of DNA in centrioles was discussed around 1970 (Granick and Gibor 1967; Pickett-Heaps 1969; Fulton 1971; Schnepf and Brown 1971).

With time, discourses favorable for the endosymbiotic hypothesis became abundant in major research journals. Raven (1970) proposed multiple and parallel endosymbiotic hypothesis, in which chloroplasts with different pigment systems were acquired independently many times. In addition, he thought that mitochondrial endosymbiosis preceded chloroplast endosymbiosis. He also suggested a situation in which mitochondria were produced from plastids, leaving the possibility of multiple origins of mitochondria. A popular scientific journal *Scientific American* published an article supporting the endosymbiotic origin of organelles (Goodenough and Levine 1970), which was soon followed by an article of Margulis (1971), which was her first popular commentary on the endosymbiotic origin of organelles, namely, chloroplasts, mitochondria, and flagella.

Cohen (1970) examined endosymbiotic origin of organelles from various aspects, and presented the two publications of Margulis, Sagan (1967) and Margulis (1970), in the final section. However, he concluded his article by negative evaluation of endosymbiotic hypothesis. In contrast to the multiple, parallel endosymbiotic hypothesis of Raven (1970), a phycologist Lee (1972) proposed a single endosymbiosis engendering cryptophytes as an ancestor of all eukaryotic algae and plants. Although he cited some contemporary papers, Lee (1972) did not cite or mention Margulis, Mereschkowsky, and other related papers on endosymbiotic origin of organelles. He changed his opinion later and stated that glaucophytes were the initial algal group originating from endosymbiosis (Lee 1980). Inspired by Sagan (1967), Raff and Mahler (1972) presented differences between bacteria and mitochondria to propose their original mechanism of evolution. Uzzell and Spolsky (1974) gave a balanced evaluation of endosymbiotic and autogenous hypotheses of organelle origin from the viewpoint of phylogenetics, and proposed a probable phylogenetic tree without supposing endosymbiosis. Because they believed endosymbiosis a kind of creation, they tried to interpret data without it, as an accumulation of mutations. As other phylogenetic trees by autogenous hypotheses, their tree placed blue-green algae at the root of eukaryotes. A precise phylogenetic tree was expected to decide which of the two hypotheses were correct. Some errors in Raven (1970) were corrected by Taylor (1974a), who was not the Taylor that proposed the SET. Interestingly, Cavalier-Smith (1975), who became later a major advocate of the endosymbiotic theory, and Bogorad (1975), who was a leader of chloroplast molecular biology at the time, both denied endosymbiotic hypothesis and supported autogenous hypothesis. Surprisingly, these two papers appeared in *Nature* and *Science*, respectively. The former stated that eukaryotes originated from blue-green algae by re-organization of membrane systems. The latter presented evidence that a single ribosome is made of subunits encoded by the chloroplast genome and those encoded by the nuclear genome, which made the author skeptical to the endosymbiotic origin of chloroplasts.

In this period 1970–1975, both endosymbiotic and autogenous hypotheses coexisted in confusion. Cohen (1973), in contrast with Cohen (1970) that was skeptical to the endosymbiosis hypothesis, cited Margulis as the first reference in the first section, listed various new data showing the similarity of chloroplasts and blue-green algae, and concluded:

> it should be clear that the apparently academic and theoretical search for organelle origins really holds many possibilities of enriching both theoretical and practical biology. (p. 444)

He indicated that important evidence for the endosymbiotic hypothesis would be (1) hybridization between chloroplast rRNA and blue-green algal genomic DNA and (2) immunological cross reaction between red algal and blue-green algal phycobiliproteins. He thought the studies on the possible relationship between chloroplasts and blue-green algae were more promising than the search for mitochondrial origins.

Taylor (1974b) named Margulis theory "Serial Endosymbiosis Theory" (SET). Note that he correctly cited three papers of Mereschkowsky, Famintzin's paper, as well as Echlin (1966) and Goksøyr (1967). In the same issue of *Taxon*, papers presented at the First International Congress of Systematic and Evolutionary Biology (ICSEB), held in Boulder, Colorado, on 11 August 1973 by the leadership of Margulis, were compiled. Taylor (1979) explained the SET in more detail. Stanier (1974) concluded that red algae resulted from endosymbiosis of blue-green algae. But he remained prudent regarding the origins of other algae.

In the early 1970s, Japanese researchers were already interested in the endosymbiotic theory. Professor Hajime Ishikawa (1940–2005) was one of them. He had a laboratory in front of our laboratory, when I was a graduate student under the supervision of Professor Norio Murata (1939–) in the University of Tokyo, Komaba. He was studying symbiotic bacteria in aphids. Ishikawa (1977) pointed out that mitochondrial rRNA was quite different from prokaryotic rRNA, while he admitted that chloroplast ribosome was of prokaryotic origin. I examined various textbooks at hand in the 1970s to re-examine the situation in Japan. In *Photosynthesis* edited by Furuya et al. (1971J) and *Introduction to Photosynthesis* by Sakae Kato (1973J), I found no reference to endosymbiotic origin of chloroplasts. In contrast, *Cell* by Shichiro Sato (1975/1979J) was an expression of strong sympathy to the endosymbiotic theory. In *Biological Membranes and Energy* by Yasuo Kagawa (1975J), similarity of bacteria and mitochondria was taken as evidence for endosymbiotic origin of mitochondria. Hakobu Nakamura (1987J) still developed a strong criticism against endosymbiotic theory of organelle origin. Tsuneyoshi Kuroiwa (2000J), leader of our Cyanidioschyzon genome project in the 2000s, presented the author's passion to mitochondria that had formed eukaryotic cells.

5.5 Development of Molecular Phylogeny and Establishment of the Endosymbiotic Theory

In the 1970s, advances in molecular phylogenetics allowed to estimate phylogeny and evolution using amino acid sequences of proteins and nucleic acid sequences of DNA and RNA. Details of molecular phylogenetic methods will be explained in Chap. 6. Here, we will overview the history of molecular phylogenetic studies that contributed to the demonstration of the endosymbiotic theory. One of the roots of molecular phylogenetics was Fitch and Margoliash (1967). The theoretical basis of molecular phylogeny was established by the neutral theory of molecular evolution by Kimura

(1968). In the 1970s, newly developed methods of molecular phylogenetics were applied to support endosymbiotic origin of organelles. Dayhoff et al. (1974) used 5S rRNA, cytochrome *c*, and ferredoxin to estimate a phylogenetic tree that was consistent with the endosymbiotic theory. Bonen and Doolittle (1975, 1976), Zablen et al. (1975), and Buetow (1976) used the fingerprint method, in which partially digested nucleic acids were catalogued to analyze 5S rRNA and showed that chloroplast rRNA was prokaryotic. Fingerprint method was the only method to analyze nucleic acids at the time, because nucleic acid sequencing was not developed yet. These authors discussed the problem by contrasting the autogenous hypothesis and the endosymbiotic hypothesis by citing Famintzin (1907), Mereschkowsky (1905a), and Margulis (1970). Bonen et al. (1977) prepared the nucleotide catalog of the 18S rRNA of wheat germ mitochondria (note that the size of mitochondrial rRNA is 18S but not 16S) by radiolabeling technique, and supported prokaryotic origin of mitochondria. They cited Margulis (1970). Schwartz and Dayhoff (1978) established the essential difference of prokaryotes and eukaryotes by molecular phylogeny. They presented a composite phylogenetic tree of 5S rRNA, ferredoxin, and cytochrome *c*, indicating a monophyletic relationship of organelles and bacteria, as evidence for the endosymbiotic origin of mitochondria and chloroplasts. Doolittle and Bonen (1981) showed close relationship between chloroplasts and cyanobacteria by phylogenetic tree based on distance matrix of 16S rRNA sequences. However, the root of the phylogenetic tree (the start point of evolution in phylogenetic tree: see Sect. 6.3) was not unambiguously determined by the lack of outgroup (this notion was introduced later). Strictly speaking in the present-day knowledge, if the root was undetermined, it was not possible to identify the monophyly of chloroplasts and cyanobacteria. Comparison of red algae and Euglena also suggested a polyphyletic relationship of chloroplasts.

The era that this chapter describes is basically until the end of the 1970s, but, as a summary of historical descriptions in Part I, we will briefly overview later development of the endosymbiotic theories. Gray and Doolittle (1982) in Dalhousie University, Halifax, Canada, were able to conclude that there was "no room for doubt" for the endosymbiotic origin of chloroplasts:

> The data discussed above, with the strong similarities in the biochemical, physiological, and morphological properties of plastids and free-living oxygenic-photosynthetic procaryotes (cyanobacteria and *Prochloron*), leave no room for doubt that the latter were ancestral to the former. (p. 9)
>
> Plastids show an obvious and close affinity to oxygenic-photosynthetic procaryotes (cyanobacteria and *Prochloron*) and appear on several grounds to be polyphyletic, thus fulfilling the criteria for the third form of "proof "outlined on at the beginning of this paper. (p. 29)

They were still prudent about the mitochondrial origin:

> The data obtained so far seem too few to allow us to conclude that criteria for any of the three sorts of "proof" outlined at the beginning of this paper have been fulfilled for mitochondria. However, the data do give hope that the criteria ultimately will be fulfilled and indicate that the best chance for fulfillment lies in further analyses of plant mitochondrial genomes. (p. 30)

Wallace (1982) agreed, based on the structures of organelle genomes (gene organization) reported by that time, the possibility of endosymbiotic origin of organelles:

> … it is now possible to propose tentative scenarios for the evolution of chlDNAs and mtDNAs. These scenarios assume an endosymbiotic origin for organelle DNAs. They also assume that the primary trend in organelle DNA evolution has been toward genome reduction, that organelle DNAs of multicellular organisms have retained primitive traits due to genome stabilization, and that the mtDNA genetic codes diverged from the universal code by a sequential series of mutations which occurred after the mitochondrial symbiont became established. (p. 227)

But he was also cautious:

> These chlDNA and mtDNA phylogenies are admittedly tentative and, in many cases, quite speculative. However, they do provide a valuable framework for organizing information and designing experiments. (p. 229)

Taylor (1987), the founder of the SET hypothesis, summarized endosymbiotic theories of mitochondria and chloroplasts, and concluded that the debate was over:

> … while the SET has acquired wide acceptance, some of its other consequences are not fully realized. (p. 13)

Its (=the SET's) predictions have been upheld by increasingly powerful molecular biological tools, and rejection can only be by denial of the value of the heart of modern biology, molecular sequencing. To the historian of science, the SET controversy offers an excellent model of the causal relationships of observations and ideas, inertia and momentum, pontification and timidity. The story is worth following: it has been fascinating so far (p. 14).

Nevertheless, the paper of Cavalier-Smith (1987) in the same issue stated a hypothesis on the endosymbiotic origin of microbody, suggesting that the debate was still going on. It is interesting to note that a perspective of history of science was used in Tayor's arguments to justify the endosymbiotic theory as in the writings of Margulis. In 1992, we can see that the situation had changed dramatically: Gray (1992) affirmed that both mitochondria and chloroplasts originated by endosymbiosis. In this way, the endosymbiotic origin of mitochondria and chloroplasts was established by the end of the twentieth century.

Before closing this chapter of historical description, we will discuss some additional points on the endosymbiotic theory.

5.6 Multiple, Parallel Endosymbiosis and Prochlorophyta

It is interesting to add an episode that reminds us of a tragic or ironical history of biology. The sensational discovery of *Prochloron* sp. in 1975 was taken as a proof of the endosymbiotic theory of Margulis (Newcomb and Pugh 1975; Lewin and Withers 1975; Lewin 1981). As described above, Margulis presupposed multiple, parallel endosymbiosis of chloroplasts. Red algae were assumed to have acquired, by endosymbiosis, a common blue-green alga that contains chlorophyll *a* and phycobiliproteins as do red algae. In contrast, the hypothetical endosymbiont that engendered green algae was supposed to have chlorophylls *a* and *b* but not phycobiliproteins. *Prochloron* sp., isolated as a prokaryotic green alga from an ascidian, was found to contain chlorophylls *a* and *b* like eukaryotic, authentic green algae. One of the discoverers, R.A. Lewin, was already acknowledged in Sagan (1967). Many decent biologists, seemingly fascinated by this young, smart, female researcher, tried to demonstrate her hypothesis. This reminds me of a recent proposal of the "STAP cell," which was completely denied in contrast with the endosymbiotic theory. The new photosynthetic prokaryote with chlorophylls *a* and *b* was highlighted as a prokaryotic green alga that became the chloroplasts of green algae. Similar prokaryotic algae named *Prochlorothrix* and *Prochlorococcus* were subsequently discovered and studied intensively.

Professor Lewin visited all over the world to convince researchers the significance of their discovery. He was invited by Professor Satoru Murakami (1933–2016) to the Komaba campus, the University of Tokyo, where I was working as a graduate student. He was a friendly, bright personality, and introduced his works under the authorship "al. et Lewin," meaning all collaborators and himself. In this way, he acknowledged his coworkers. As indicated by this fact, he was not ambitious at all. Rather, he was very prudent, asking the audience whether the evidence was enough to conclude the hypothesis of prokaryotic green algae. Nevertheless, many researchers were excited, and we shared the sentiment that this was a big discovery.

The situation changed gradually. Norio Murata and myself showed, in *Prochloron* sp., the presence of a glucolipid, monoglucosyl diacylglycerol (Murata and Sato 1983), which is a common constituent in all cyanobacteria (see Sato 2015) but not green algae. *Prochlorococcus marinus* CCMP 1375 was found to contain chlorophylls *a* and *b* as well as phycobiliproteins (Hess et al. 1996). Phylogenetic analysis showed that *Prochlorococcus* was included in a cyanobacterial subclade (Palenik and Haselkorn 1992; Urbach et al. 1992), now called α-cyanobacteria with marine species of unicellular cyanobacteria. Phylogenetic analysis using 16S rRNA sequences clearly indicated that various "prokaryotic green algae" were included within the cyanobacterial clade, and that they were not monophyletic. In other words, each prokaryotic green alga was placed in a different subclade of cyanobacteria. Genomic analysis confirmed cyanobacterial nature of prokaryotic green algae (Dufresne et al. 2003; Rocap er al. 2003). Recent results of phylogenetic analyses will be shown in Chap. 6.

The hypothesis of prokaryotic green algae could be viewed as an abominable by-product of the endosymbiotic theory of Lynn Margulis. Nevertheless, the studies on prokaryotic green algae were not useless. We now know that picoplanktons (very small planktons, about 1 μm in diameter) including *Prochlorococcus* are major primary producers in the ocean, responsible for about a half of total marine productivity. Diatoms were traditionally considered important producers in marine ecosystems, but picoplanktons had escaped from detection because of their small sizes.

5.7 Generalization of the Term "Endosymbiosis"

In contemporary literature, the word endosymbiosis is routinely used to mean the origin of chloroplasts and mitochondria, but this is not an old custom. A measure of usage of this word (endosymbiosis or endosymbiotic) was obtained by searching the PubMed database of the National Center of Biotechnology Information (NCBI), and listed in Table 5.6.

The term endosymbiosis was originally a rare word used in specific fields such as ecology, but beginning from the 1970s, it became increasingly widely used, and the usage exploded from the mid-1990s, which corresponded the time when genome sequencing of bacteria, plants, and animals became popular in biological studies. This suggests that the development in genomic science contributed to the spread of the term endosymbiosis. Owing to their relatively small sizes, many reports of chloroplast genomes were published from the time. Each sequencing paper stated that chloroplasts originated from an ancient cyanobacterial endosymbiont. Subsequently, papers reporting nuclear genome sequences classified identified genes using information on whether they encode organellar proteins. Homology search has been performed for this purpose to identify homologs of bacterial or cyanobacterial proteins. Then, researchers understood that these homologs originate from endosymbiosis. Unfortunately, many bioinformatics researchers of the time had little biology education, but used introductory textbook knowledge to apply the notion of endosymbiosis to gene annotation. In my impression on the atmosphere of the time, this was a plausible reason explaining the spread of the term endosymbiosis. In this way, the notion "organelles = endosymbiosis" was gradually settled in genomic science, without examining detailed situation of individual endosymbiosis events and the evidence for the endosymbiotic origin of organelles. The 1970 book of Margulis was a good reference to cite in this situation.

Table 5.6 Time evolution of the number of articles using endosymbiosis or endosymbiotic

Period	Number of articles using endosymbiosis or endosymbiotic	Total number of articles	Per mille
1960–1969	6	1,632,480	0.004
1970–1979	45	2,455,754	0.018
1980–1989	78	3,325,127	0.024
1990–1999	258	4,429,951	0.058
2000–2009	1018	6,513,308	0.156
2010–2016	1343	7,023,869	0.191

Adapted from Sato (2017)

It is difficult to survey book citations, but the website Google Scholar helps us to count citations. According to this, the citation of Margulis (1970) remained 260–310 per year during the 1970s, 1980s, and 1990s, but increased to 450 in the 2000s and 2010s. This suggests increase in the citation in the genomic era, although the change was not as dramatic as the change in paper citations.

5.8 Developments in Genetics and Endosymbiotic Hypotheses: DNA-Centrism

Developments in endosymbiotic hypotheses curiously correspond to developments in genetics (Table 5.7). In the early twentieth century, researchers who were not satisfied with the Darwinian theory of evolution proposed endosymbiotic theories. Capability of genetics in explaining evolution is likely to reflect in the plausibility or validity of endosymbiotic explanation of organelle origin. Therefore, it seems interesting to find a relationship between the developments in genetic knowledge and the advances in endosymbiotic hypotheses. This is, in essence, closely related to the fact that the major evidence for the endosymbiotic origin of organelles resides in the presence of organellar DNA. Discovery of cytoplasmic inheritance and cytoplasmic genetic elements certainly enhanced the image of endosymbiont living inside the cell, independently from the cell nucleus. In contrast, researches in photosynthesis clearly indicated that all photosynthesis with oxygen evolution has essentially identical mechanisms, thus establishing functional unity of all photosynthetic organisms. This had a pivotal importance in endosymbiotic hypotheses of plastids: namely, identical mechanism of photosynthesis strongly supported a phylogenetic relationship between cyanobacteria and plastids. One idea focusing on the plastid was the autogenous hypothesis or "botanical myth," whereas another idea focusing on the eukaryotic cell was the endosymbiotic theory. Both Mereschkowsky and Margulis considered that plastids with different pigment systems originated from

Table 5.7 Relationship between the developments in genetics and the advancements in endosymbiotic hypotheses on the origin of plastids

Period	Genetics	Endosymbiotic hypotheses
1900s	Rediscovery of Mendel's law (birth of genetics), discovery of cytoplasmic inheritance	Mereschkowsky's endosymbiotic hypothesis
1960s	Birth of molecular biology Discovery of cytoplasmic genetic factors	Endosymbiotic hypotheses based on cytoplasmic inheritance, organellar DNA, and organellar protein synthesis
1980–1990s	Extensive DNA sequencing, birth of genome science	Endosymbiotic hypotheses based on genome phylogeny, proposal of secondary endosymbiosis
2000–2010s	Post-genomic analysis	Re-recognition of the discontinuity of cyanobacteria and chloroplasts

endosymbiosis of different prokaryotic algae, which turned out to be an error in the end. Nevertheless, this hypothesis resulted in discoveries of independent primary endosymbioses and secondary endosymbioses, suggesting that it was a fruitful error. Perhaps, we expect to find some other examples of plastid endosymbiosis in the future. A major current consensus affirms a single endosymbiotic origin of mitochondria, but we have to understand that mitochondrial DNA is quite diverse. Current evidence for the single origin of mitochondria might not be strong enough to completely reject a counter hypothesis of multiple origins. Various organelles such as hydrogenosomes acting under anaerobic conditions in certain eukaryotes are now considered as related to mitochondria (see, for example, Martin 2017). These organelles do not contain DNA, excluding the possibility of identifying their origin by DNA phylogeny. These various organelles could have originated separately. We still do not have a definite answer on the origin of mitochondria.

A major reason that the endosymbiotic origin of organelles is widely accepted in the late twentieth century was the DNA-centered thoughts in biology. The success of molecular biology in the 1960s and 1970s was reinforced by the DNA sequencing technologies in the 1980s. Because DNA determines all the characters and behaviors of a cell (or an organism), the origin of an organism, a cell, and an organelle is identified by the origin of DNA. As an extension of the success of universal phylogeny of all organisms using 16S rRNA, the origin of organelles had to be known by the 16S rRNA. Bacterial phylogeny can be determined by ribosomal RNA. People ignored that the situation in organelles is complicated. Even now, many researchers believe that the origins of plastids and mitochondria are identified by phylogeny of ribosomal RNA. Origin of organelle = origin of organellar DNA = origin of rRNA. This is a simplified scheme of thinking.

5.9 Margulis' Secret of "Success"

5.9.1 Political Situation "Cold War"

In the 1960s, a tension existed between the Soviet Union (communist state before Russian Federation) and the United State of America, as culminated in the Cuban Crisis with a serious threat of the Third World War. The long-lasting tension was called Cold War, in which all countries were divided into two groups, one including communist or socialist states headed by the Soviet Union (Eastern countries) and another including Western Europe, Japan, South Korea, and other countries that were protected by the overwhelming power of the United States (Western countries). Under such situation, we can imagine how it was difficult for an American scientist to recognize the work of Russian scientists, even though they lived in the Russian Empire before the Russian Revolution. We must remind ourselves that many Russian biologists were involved in the discourses of endosymbiosis or symbiosis. This was certainly fostered by the social background in which an idea of mutualism (or social symbiosis) prevailed, perhaps related to the socialist utopia. As indicated by the ideals of English and French Revolutions, Western culture emphasized the value of liberty, but did not have affinity to the idea of symbiosis. The United States of America was a country symbolizing liberty, supported by the number-one economical power in booming economy after the Second World War.

Biological notion of symbiosis necessarily accompanies an anthropomorphic image of social mutualism. I have an impression that many reports of ecological research use a technique of anthropomorphism to better convince the audience. This was not the case in molecular or cellular biology. The rise of molecular biology prompted to understand everything in the cell in a mechanistic way. In the 1960s, there was a deep groove between conventional biologists and molecular biologists. The biologists that supported the endosymbiotic hypothesis tried to import the macroscopic notion of symbiosis into the cell. As Margulis repeated, the notion of holism, namely, that an organism is a unity as a whole which is more than the sum of parts, was a revival of the old biological trend, and at the same time, a strong antithesis against the reductionism promoted by molecular biology. Nevertheless, the biologists proposing endosymbiotic hypotheses were not allowed (or did not dare) to identify Russia as the origin of their symbiotic idea, because they had a risk of being condemned as collaborators of "dangerous" communism. The Cold War must have been a reason that Margulis tried to present her narrative entirely as her original. This contrasted with her activities as a presenter of Russian studies after the collapse of the Soviet Union, such as the publication of translation of Khakhina's book. This resulted in an ironical situation in which Margulis was praised for her discovery of ancient Russian biologists (Lazcano and Perera 2017). This is not a correct evaluation of Margulis' contribution (Sato 2017), as I explained above.

5.9.2 *Multifaceted Knowledge*

We are surprised to know that Margulis wrote her first paper proposing the endosymbiotic theory of eukaryote origin at the age of 29 and that she wrote a voluminous book 3 years later. Do we find such a productive postdoc in the laboratories in our departments? Some of them might be good at experiments, performing elaborately planned measurements using finely manipulated organisms. Nevertheless, do some of them scrutinize literatures on various different disciplines to propose an original view on biological world? Perhaps we can find some young, bright scientists actively working in their original researches exploiting their wide background knowledge, but Margulis was blessed by exceptional erudition with multifaceted knowledge. As explained earlier, she must have obtained help by Carl Sagan, her first husband, in obtaining information on Earth science, but we imagine how she made hard efforts in learning: starting from cytology as a basis, she learned Earth sciences including Earth history, history of life, and fossil records on the one hand, and biological sciences including molecular biology, biochemistry, and physiology on the other hand. Margulis was certainly young enough to master such wide expertise with her bright intelligence, without disturbed by too much knowledge of a single discipline. A fresh Ph.D. without knowledge bias might be able to absorb any information. Young Margulis, or Lynn Alexander (maiden name), must have been an exceptionally talented feminine scientist.

5.9.3 *Taste of Philosophy of Science*

In addition to scientific knowledge, Margulis was able to use philosophy of science in supporting her proposal. After the bestseller *Silent Spring* by Rachel Carson (1962/2002), which pointed out serious environmental pollution caused by industrialization during the post-war economy, doubt about scientific knowledge was growing in industrialized countries. In Japan, environmental pollution problems such as the spread of methyl mercury in Minamata Bay were closed up. As a high-school student, I was trying to measure air pollutants in the late 1960s. At the time, the sky was white, not blue, in Tokyo city, perhaps due to photochemical smog. Some crossings famous for dirty air, such as Chitose-bashi and Ohara in Tokyo, were covered by gray gas. Note that the situation has been surprisingly improved, fortunately.

Under this situation, ideas doubting the value of science itself became apparent. Already in the 1950s, re-evaluation of the scientific logic and the theoretical grounds of scientific knowledge were beginning in philosophy of science. Karl Popper (1959/2002) considered that "falsifiability" was the best criterion of a good scientific hypothesis. This notion affected not only physics but also biology, as cited in *Le hasard et la nécessité* by Jacques Monod (1970: English translation *Chance and Necessity* in 1971). Thomas Kuhn presented a famous notion of paradigm in *The Structure of Scientific Revolution* (Kuhn 1962/1970), causing a worldwide shock. This notion was applied to a radical, systematic change in the manner of scientific thinking. The term paradigm is still used in various different situations, such as politics and culture. Margulis (1981/1993) also used this term as a section title ("Paradigm Symbiosis" in Chap. 7) to contrast "symbiotic paradigm" and "direct-filiation paradigm." The new trend in philosophy of science of the time preferred to find value in only selected scientific knowledge that satisfied certain conditions, rather than to accept uncritically all scientific knowledge.

The discourses of Margulis reflected the notion "falsifiability." Although she did not cite Popper, she repeatedly used "verifiable" and "testable" in this meaning, beginning from her 1967 paper (Sagan 1967; Margulis 1970). She claimed that her hypothesis was not a fancy idea but an authentic scientific hypothesis that can be verified by experiments. This attitude became explicit in some theoretical papers discussing the falsifiability of her hypothesis (Margulis 1975; Margulis and Barmudes 1985). In these papers, possible consequences of her hypothesis were presented for each of the three endosymbiotic scenarios, namely, the origin of mitochondria, chloroplasts, and undulipodia (the 9+2 structures). She also suggested probable experiments that were presumed to verify them.

Now, we have a serious question. The notion of falsifiability that Popper proposed was originally thought to be applied for hypotheses on physical laws, like this: if hypothesis A is correct, then consequence B will be true, and consequence C will also be true, and so on. If any of the probable consequences is false, then the original hypothesis A must be false. Hypothesis A should be either rejected or modified. The hypothesis of Margulis was different. It was itself a complex hypothesis, consisting of three different endosymbiotic events, namely, mitochondria, chloroplasts, and (9+2) structures. It also included a hypothesis that mitotic apparatus originated from flagella. A large number of consequences (or predictions) were deduced from these component hypotheses. It was quite improbable that all of them could be verified by experiments. In fact, most of the experiments proposed to verify her hypotheses gave negative results. For example, the endosymbiotic origin of chloroplasts, which seemed to be more or less valid at that time, must have been judged false, because Margulis considered that the best evidence for endosymbiotic origin of chloroplasts was in vitro culture of isolated chloroplasts, which is obviously impossible in the current knowledge. The notion of falsifiability of Popper, if rigorously applied, meant that only a single falsification of a possible consequence was enough to reject the initial hypothesis. Margulis, however, did not change her whole narrative, even though many of her predictions were rejected, including the famous spirochete origin of flagella. On the contrary, she was proud that two of her endosymbiotic stories (mitochondria and chloroplasts) were scientifically verified (Margulis 1998, p. 38: actual citation will be in Sect. 9.2). It seems that the argument of falsifiability that Margulis used repeatedly was, in fact, a very well structured trap or rhetoric, emphasizing her correctness.

5.9.4 Young, Feminine Researcher Embodying Scientific Advances

Nevertheless, Margulis' proposal was strong enough to move many important scientists of the time. Many principal microbiologists of the time, such as Roger Stanier, acted in support of her hypothesis, maybe because her proposal was rooted in her exceptional knowledge on Earth science, fossil records, and origin of life that was obtained from top-level scientists.

The most famous and perhaps the first young, feminine scientist was Marie Curie (1867–1934). Coming from Poland to Paris, she married a physicist Pierre who performed less conspicuous study. With his help, she worked very hard for her Ph.D. study, purifying minute amounts of polonium and radium from hundreds of kilograms of rocks, and obtained the first evidence for radioactivity (Curie 1899). At the time around 1900, women had difficulty in pursuing research, but fortunately she had some kind supporters who arranged two Nobel Prizes, one with discovery of radioactivity and the other with discovery of radium. Marie put forward a curious hypothesis including radioactive decay of heavy metal, reminiscent of alchemy, but

Rutherford presented rapidly a decisive demonstration of her hypothesis. Therefore, in the case of physics revolution in the 1900s, many technical advances simultaneously and cooperatively worked for new discoveries.

In the 1970s, on the other hand, biological revolution was driven by the advances in molecular biology. As represented by the famous painting of Delacroix "La Liberté guidant le peuple" (Liberty leading the people), woman is the symbol of a revolution. In this sense, no one doubted that a woman researcher like Jeanne d'Arc (a girl leading the French army to win England at the end of the Hundred Years' War in the fifteenth century) appeared in the biological revolution. In fact, Linda Bonen (Professor in Ottawa University) and Margaret Dayhoff (1925–1983) performed molecular phylogeny suggesting endosymbiotic origin of chloroplasts. Ursula Goodenough (Professor in Washington University), who wrote a commentary on the endosymbiotic theory in 1970, became a leader of *Chlamydomonas* genetics. Ruth Sager, a rival of Lynn Margulis at the time, isolated and characterized chloroplast DNA of *Chlamydomonas*. The "success" of Lynn Margulis was a part of these leading works by feminine biologists.

5.10 Summary of Part I

We analyzed the one-century history of endosymbiotic theories of organelle origin in Part I. I presented two major stars of endosymbiotic hypotheses, Mereschkowsky and Margulis. The former proposed the initial and original form of endosymbiotic hypothesis. Nevertheless, during the process in which endosymbiotic origin of organelles was invented, discussed, and established, not only biological revolution including the rise of molecular biology and innovation in genetics, cytology, and biochemistry, but also advances in Earth science, paleontology, and theory on the origin of life worked in concert. We found many different reasons why Margulis became the champion of the endosymbiotic theory in the end: A major reason was obviously her talent, but there were several additional, peripheral conditions that favored her success. In my view, the initial idea of Margulis on the origin of eukaryotic cells tried to explain the evolution of mitosis by excluding mitochondria and chloroplasts. Endosymbiotic hypothesis was necessary for her to remove these organelles from the main history of eukaryotic cells. In other words, she was not seriously interested in the real origin of these organelles. However, the original hypothesis of Margulis was denied later, while she became a symbolic advocate of the endosymbiotic theory. This was an impressive change of attitude (Sato 2017), which was also a part of her talent.

The separation of organelles from the eukaryotic cell, which was never accepted by Paul Buchner, has a political advantage to the development of biology. In the 1980s, studies on the cell division and cell cycle (mostly in animals and human) progressed extensively in Japan by the initiative of the late Prime Minister Nakasone to fight against cancer, and this movement continued until the end of the twentieth century. The same was true in many other countries. As a result, mechanisms of many important biological phenomena were unraveled, such as various intracellular signal transduction systems, equal partitioning of chromosomes, and apoptosis (programmed cell death). Meanwhile, studies on chloroplasts and cyanobacteria survived in the form of genome projects to reveal evolution of photosynthetic organisms. It was the separation of organelles from the main cell biology that kept the independence of chloroplast molecular biology. From the year 2000 onwards, all biological studies shifted to genomic studies with a common language of genome. Until this large shift in trends, studies on chloroplasts and cyanobacteria gained steady progress without serious interference from cell biology. We could attribute this fortunate situation to the strong influence of Lynn Margulis.

Taylor (1987) summarized the contribution of Margulis to the endosymbiotic theory:

> I have previously attributed the serious revival of interest largely to the conclusive demonstration of DNA in plastids and mitochondria in the 1960s. Another critical factor was Margulis' active and provocative espousal of the hypotheses in a synthetic form, within a broad geological context, in a germinal book in 1970. It is one of the ironies of scientific history that her own original contribution to the question, i.e., the origin of "9+2″ microtubular structures by a further symbiosis, has not been widely accepted. However, her book represents the first synthesis of the separate hypotheses for each organelle into a general theory of eukaryotic cell evolution (unnamed and not presented as such at the time), marshalling the available supportive evidence and indicating predictions that could be, and were, tested by cell biologists. (Taylor 1987, p. 5)

In Part I, I explained that Margulis was not the first proponent of the endosymbiotic theory. Mereschkowsky was the first to propose endosymbiotic origin of chloroplasts, whereas Goksøyr presented a scheme of serial endosymbiotic hypothesis comparable to the currently accepted one. Nevertheless, it is necessary to reveal the original idea of Margulis by carefully reading her writings, and this is a sincere attitude towards the works of Margulis. In addition, I have pointed out her many advantages in the progress in organelle biology in this section. I hope this could be a basis of historical studies on modern biology of organelles.

Part II
Current Perspectives

We analyzed historical aspects of endosymbiotic hypotheses of organelle origin. Part I is therefore devoted to the history of biology, whereas Part II describes current status of scientific knowledge. In this part, I will explain contemporary understandings of organelle endosymbiosis, mainly based on the results of phylogenetic analysis performed by my colleagues and myself to renew conventional idea of the endosymbiotic theory.

Chapter 6
Phylogenetic Evidence for the Endosymbiotic Origin of Organelles

In this chapter, we will examine various pieces of evidence for the endosymbiotic origin of plastids. As we found in Part I, similarities of various characters of plastids and cyanobacteria have been taken as evidence for the cyanobacterial origin of plastids. More conclusive evidence is now provided by molecular phylogenetics. In this chapter, I will explain basic theories of molecular phylogenetic analysis for nonspecialists, and then I will present evidence that supports cyanobacterial origin of plastids. Plastids will be the major topic, because more supportive evidence can be found for the endosymbiotic origin of plastids than the origin of mitochondria, which is much more complicated by the presence of anaerobic, homologous organelles such as hydrogenosomes devoid of DNA. Indeed, I do not find definitive evidence, either historical or contemporary, for the endosymbiotic origin of mitochondria except for the phylogeny of mitochondrial rRNA. This point will be briefly explained later.

6.1 Phylogenetic Tree and Molecular Phylogeny

What kind of evidence exists for the cyanobacterial origin of plastids? We will consider this problem from the current status of biological knowledge, which must be different from the knowledge in the 1970s. In those days, enumeration of similarities of cyanobacteria and plastids was the major arguments for the cyanobacterial origin of plastids (Chap. 5, Table 5.5). Major similar characters included oxygenic photosynthesis, presence of chlorophylls, prokaryotic ribosomes, and presence of DNA. Table 6.8 summarizes the similarities from the contemporary standpoint. Three types of characters are distinguished, namely, substance, structure, and function, because complex structure and function cannot be explained simply by the presence of specific substances.

Now, detailed phylogenetic analysis enables us to judge objectively whether we can attribute the origin of chloroplast genes and proteins to cyanobacteria. In this sense, molecular phylogenetics became the decisive evidence for (or against) the

N. Sato, *Endosymbiotic Theories of Organelles Revisited*,
https://doi.org/10.1007/978-981-15-1161-5_6

Table 6.8 Similarities of cyanobacteria and chloroplasts

Classification	Item	Cyanobacteria	Chloroplasts
Substance	Chlorophyll *a*	Present	Present
	Chlorophyll *b*	Present in *Prochlorococcus*	Present in green algae/plants
	Glycolipids	MGDG, DGDG	MGDG, DGDG
	DNA	Present	Present
	Ribosomal RNA	Prokaryotic	Prokaryotic
	Enzymes/proteins	Share many protein families of monophyletic origin	
Structure	Thylakoid membrane	Present except in *Gloeobacter*	Grana are present in green algae/plants
	Phycobilisome	Present except in *Prochlorococcus*	Present in red algae
	Envelope membrane	Outer and inner membranes	Outer and inner membranes
	Peptidoglycan	Thick and conspicuous	Conspicuous in glaucophytes Thin in mosses and ferns
Function	Reproduction	Binary fission	Binary fission
	Protein synthesis	Prokaryotic	Prokaryotic
	Photosynthesis	Oxygen evolution	Oxygen evolution

Glycolipids and peptidoglycan are explained in the text. Chlorophyll *a* is present in the reaction center of photosystems, whereas chlorophyll *b* is an accessory pigment. Phycobiliproteins are also accessory pigments that form phycobilisomes. *Prochlorococcus* cells do not possess phycobilisomes but light-harvesting proteins containing chlorophyll *b*

endosymbiotic (or cyanobacterial) origin of plastids. Many studies, however, tended to hastily conclude cyanobacterial origin of a plastid protein by just finding homologous proteins in cyanobacteria and chloroplasts. In my view, molecular biologists are very prudent in analyzing the function of a gene, but they tend to conclude a phylogenetic relationship quite easily from partial evidence. In fact, many plastid proteins are not direct descendants of cyanobacterial homologs. In this case, we cannot conclude that the plastid proteins originate from cyanobacteria.

Phylogenetic inference involves a complicated logic as explained in the next section. We cannot identify the origin of a protein just by finding homologs in bacteria. As described in Sect. 1.6, it is a common, easy practice to suppose historical continuity of cyanobacteria and plastids, based on their similarities. Nevertheless, phylogenetic analysis is mandatory to confirm the continuity. As we will find later, many similar characters are results of unrelated mechanisms or enzymes, and thus evidence for the discontinuity of plastids and cyanobacteria.

I shall begin by explaining what a phylogenetic tree is. Originally, a phylogenetic tree graphically represented Darwin's idea that all organisms are descendants of a common ancestor. In early days, biologists (or naturalists) compared various characters of organisms and considered that organisms sharing a common character belonged to an identical clade. Using various different characters as markers, they found a phylogenetic relationship in which primitive organisms branched into more complex organisms having more complex morphology and function. Namely, a common ancestor evolved into plants, animals, and monera (primitive organisms).

In animals, diploblastic and triploblastic animals are distinguished from organisms with simpler morphology, and triploblastic animals are further classified into Protostome and Deuterostomia. It was quite reasonable to assume that classification by similarity represented phylogenetic relationship of organisms. Famous phylogenetic trees of Haeckel (Fig. 6.18) were drawn in this way. Classical phylogenetic trees were, therefore, images embodying the knowledge within the brain of erudite, insightful taxonomists.

This type of phylogenetics persisted until the mid-1980s. However, it was dependent on personal thinking of taxonomists and became no longer supported by molecular biologists. Molecular phylogenetics appeared as a novel technique that rapidly replaced old phylogenetics. It consists of a method of inferring phylogenetic relationship by comparing nucleic acid sequences of DNA or amino acid sequences of proteins. This new method, initially introduced in the 1970s, dramatically developed from the 1990s, under the influence of parallel, rapid development of computer technology and informatics, and became a standard methodology in biology now. Currently, many web sites provide simple interfaces of phylogenetic analysis for experimental biologists who are not familiar with phylogenetic methods.

The basic principle of molecular phylogenetic analysis is not fundamentally different from the method used by conventional phylogenetics based on similarity of characters. In molecular phylogenetics, we compare homologous nucleotide sequences or amino acid sequences, and focus on differences between organisms. If there are more differences in the homologous sequences of two organisms, we consider that they diverged at an older time. In conventional phylogenetics, similarities were used to classify organisms. In molecular phylogeny, differences were used to infer evolutionary distances. This was not possible (or at least very difficult) in conventional phylogeny. Molecular phylogenetic analysis systematically quantifies sequence differences, which are used to estimate evolutionary distances.

Take a simple example. Human and chimpanzee diverged about 7 million years ago from their common origin, and are very closely related (at least in biological terms). In contrast, dog and cockroach diverged from human in older times, in this order. In principle, all organisms must have diverged from a single organism. The fact (rather than a hypothesis) that even cockroach share a common origin with human might shock some people! We take an example: a subunit of mitochondrial NADH dehydrogenase (ND1) with 318 amino acid residues. We count a small number (6%) of sequence differences between human and chimpanzee ND1s. We count more differences (24%) between human and dog sequences. Between human and cockroach, much more positions in the sequences (50%) are different except functionally very important ones.

Sequence differences can be easily counted by a computer. This is rapid and can avoid human errors. For this purpose, many computer programs have been developed. They automatically search homologous sequences (either amino acids or nucleic acids) in many different organisms, align and compare them, and then estimate their phylogenetic relationship. A method used in early days (parsimony method) inferred a phylogenetic tree that has a minimal number of sequence changes. Distance-based methods such as Neighbor-Joining (NJ) method became

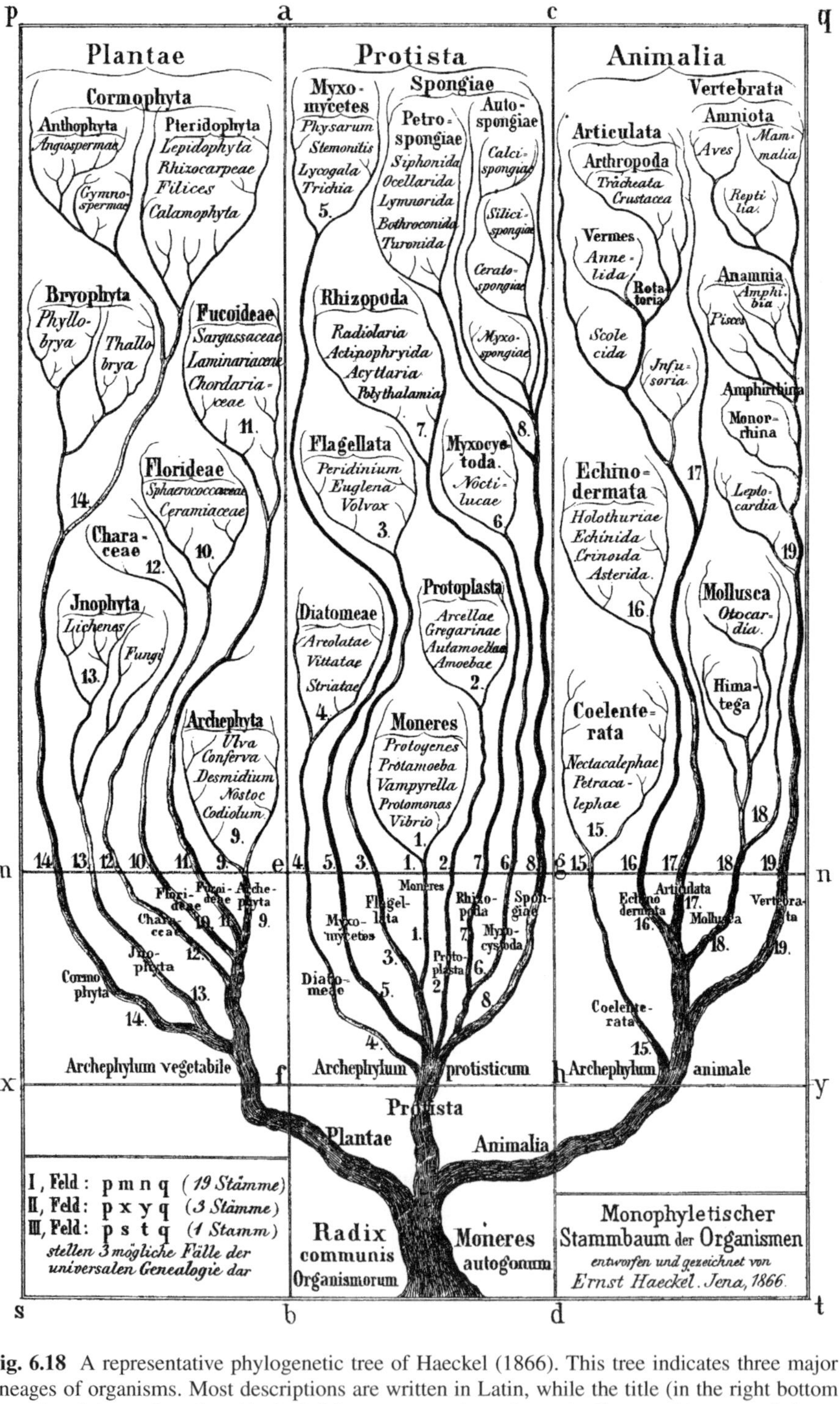

Fig. 6.18 A representative phylogenetic tree of Haeckel (1866). This tree indicates three major lineages of organisms. Most descriptions are written in Latin, while the title (in the right bottom corner) and the explanations (in the left bottom corner) are shown in German. Lineages of plants, protists, and animals are presented from left to right

Table 6.9 Steps of phylogenetic tree construction with representative software or services

Step	Service or software	URL
Homology search and sequence acquisition	NCBI (National Center for Biotechnology Information)	http://www.ncbi.nlm.nih.gov/
	EBI (The European Bioinformatics Institute)	http://www.ebi.ac.uk/
Multiple alignment	Clustal W (Clustal X)	http://bips.u-strasbg.fr/fr/Documentation/ClustalX/ ftp://ftp.ebi.ac.uk/pub/software/clustalw2
	Muscle	http://www.drive5.com/muscle/
	MAFFT	http://align.bmr.kyushu-u.ac.jp/mafft/software/
Estimation of phylogenetic relationship	MEGA	http://www.megasoftware.net/
	MrBayes	http://mrbayes.csit.fsu.edu/
	PhyML	http://www.atgc-montpellier.fr/phyml/
Drawing phylogenetic tree	FigTree	http://tree.bio.ed.ac.uk/software/figtree/
Homologous protein clusters	Gclust	http://gclust.c.u-tokyo.ac.jp/

popular for constructing large trees with many taxa. Advances in computer power in recent two decades enabled to infer “an evolutionary model” by solving a complex set of differential equations, with many parameters, such as branching pattern and replacement rates of individual bases in various branches. Many recent programs implementing the Maximum Likelihood (ML) method repeatedly calculate the parameters to fit the actual sequence data. The Bayesian inference (BI) method uses a further longer chain of calculation to achieve convergence.

6.2 Actual Methods of Molecular Phylogenetic Analysis

Here I briefly describe a flow of data analysis in molecular phylogeny for nonspecialists. Table 6.9 summarizes the analytical steps, with information sources and computer programs that I usually use. The first step in phylogenetic analysis is to obtain many homologous sequences (DNA or protein) by homology search. We first need a seed sequence of a protein family that we are trying to analyze which we get from the databases in the web sites in National Center for Biotechnology Information (NCBI) or The European Bioinformatics Institute (EBI). We may get sequences by sequence annotation, but note that not all related sequences are annotated correctly. To obtain all (or sufficiently many) homologous sequences, we use homology search. “BLAST” is the representative, standard software for this purpose. We may use the BLAST service in NCBI or EBI to search all necessary sequences.

Then, we have to align the sequences. This manipulation is called multiple alignment, in which corresponding bases or amino acids are identified in the collected sequences. Various computer programs are available for this purpose. Clustal X is a classical one that has been widely used. I have integrated Muscle in my pipeline for data processing for a long time. MAFFT is also a common program that integrates various programs within a package.

The next step is to estimate phylogenetic relationship. Neighbor-Joining (NJ) method was used widely before. Implementation of more sophisticated methods requiring a higher computational power became feasible with desktop computers of personal use. MEGA has been a representative program for the NJ method, but its recent version can cope with the ML method. Calculation by the ML method is implemented in various programs, such as RAxML and PhyML (see Table 6.9). The BI method that requires a very long computation time with parallel processing is usually used in supercomputers or high-power servers. Representative programs are MrBayes (Table 6.9) and PhyloBayes. We have to choose a proper method depending on sequence relatedness. For example, phylogenetic relationship between various human individuals originating from various regions or countries can be estimated by just comparing nucleotide changes or insertions/deletions. The NJ method or parsimony method are suited for this purpose. In contrast, phylogenetic estimations including many different species, namely, estimations over a long span of evolution, such as those important in this book, must be performed by the ML method or BI method, which includes many parameters considering the variation of evolutionary rates in different branches. In a recent symposium (2017) in the Society of Evolutionary Studies, Japan, Professor Naruya Saito, one of the inventors of the NJ method, emphasized that the NJ method is still useful in particular purposes.

Now we have the result of phylogenetic estimation, which are still a set of numbers in a file. To draw a phylogenetic tree, we need another type of software. Various tree-drawing programs exist, but I currently use FigTree, which is easy to use and provides a number of options. Except for the drawing step, all the calculations are performed on the UNIX platform such as macOS and Linux. MEGA was first developed for the Windows OS (operating system), but is now available on other platforms.

It is possible to collect homologous sequences manually one by one, but it is tedious and prone to errors. In my practice, in fact, I use a special database called Gclust including homologous protein clusters, which have been prepared automatically by clustering all proteins in a predefined set of organisms, based on the results of all-against-all BLAST searches. Homologous clusters are generated by other programs such as OrthoMCL, but Gclust software (Sato 2009) that I developed is specialized in properly clustering organellar proteins with bacterial homologues by recognizing transit peptides. Each chloroplast protein encoded by the nuclear genome has a long stretch of sequence located in its *N*-terminus (transit peptide), and thus fairly longer than the corresponding bacterial (or cyanobacterial) proteins. This difference sometimes prevents computer software from correctly recognizing homologous proteins of chloroplasts and cyanobacteria. Gclust removes the transit peptides to produce a protein cluster encompassing organelles and bacteria. In addition, differences in homology (or evolutionary rate) in different protein families are taken into consideration in forming clusters. Representative datasets are publicly available in the Gclust server. Most of the phylogenetic trees presented in this book are prepared by using the protein clusters produced by the Gclust software. The processes for preparation of phylogenetic trees are mostly automated. This is, so to speak, a special setup for attacking the problem related to the endosymbiotic theory.

6.3 Molecular Phylogenetic Evidence for the Endosymbiotic Origin of Chloroplasts

We are now ready to examine molecular phylogenetic evidence for the endosymbiotic origin of organelles. In an ideal situation, an identical phylogenetic tree will be estimated with different families of proteins or genes. Figure 6.19 shows a phylogenetic tree of prokaryotes using concatenated (i.e., tandemly linked) sequences of 48 proteins shared by various prokaryotes. Taking Archaea as an outgroup, proteobacteria (consisting of α, β, and γ subgroups), actinobacteria, firmicutes, and cyanobacteria are diversified. As prokaryotic organisms, in general, lack visible traits even under a microscope, metabolic traits, such as gelatin digestion, have been used to classify these microorganisms. But biochemical or metabolic characters were not enough for unequivocal taxonomy. Molecular genetic data give, in contrast, a good, objective, phylogenetic relationship.

Let me explain the way of interpreting phylogenetic tree for general readers. The phylogenetic trees shown in this book are composed of vertical and horizontal lines. Each horizontal line shows evolutionary distance, whereas the length of any vertical line is arbitrary. If the tree is placed in the vertical arrangement like the phylogenetic trees of Haeckel, the meanings of vertical and horizontal lines are exchanged. Note that, in all phylogenetic trees presented in this book (except Haeckel's tree), evolution begins at the left end, and proceeds to the right. The scale of evolutionary distance is shown above in Fig. 6.19. The distance "0.2" means that each amino acid residue mutated 0.2 times on average during the time corresponding to this length. The total evolutionary distance from the left end to the right end is larger than 1.0. This should be understood that an identical position of amino acid sequence mutated more than once. In all currently used methods of phylogenetic inference, evolutionary distance is corrected for multiple mutations, and becomes longer than the original length estimated from simple counting of mutations. In addition, evolutionary rate of each branch may be different. Accordingly, some branches become longer than the others, and this variation in length depends on phylogenetic methods or software.

Phylogenetic trees constructed as above are not "rooted," because there is no root, or the starting point of evolution. In principle, it is not possible to identify the root by the phylogenetic estimation based on comparison of gene or protein sequences. The calculation involved in phylogenetic estimation gives estimation of evolutionary distances between the sequences and branching pattern, but the point

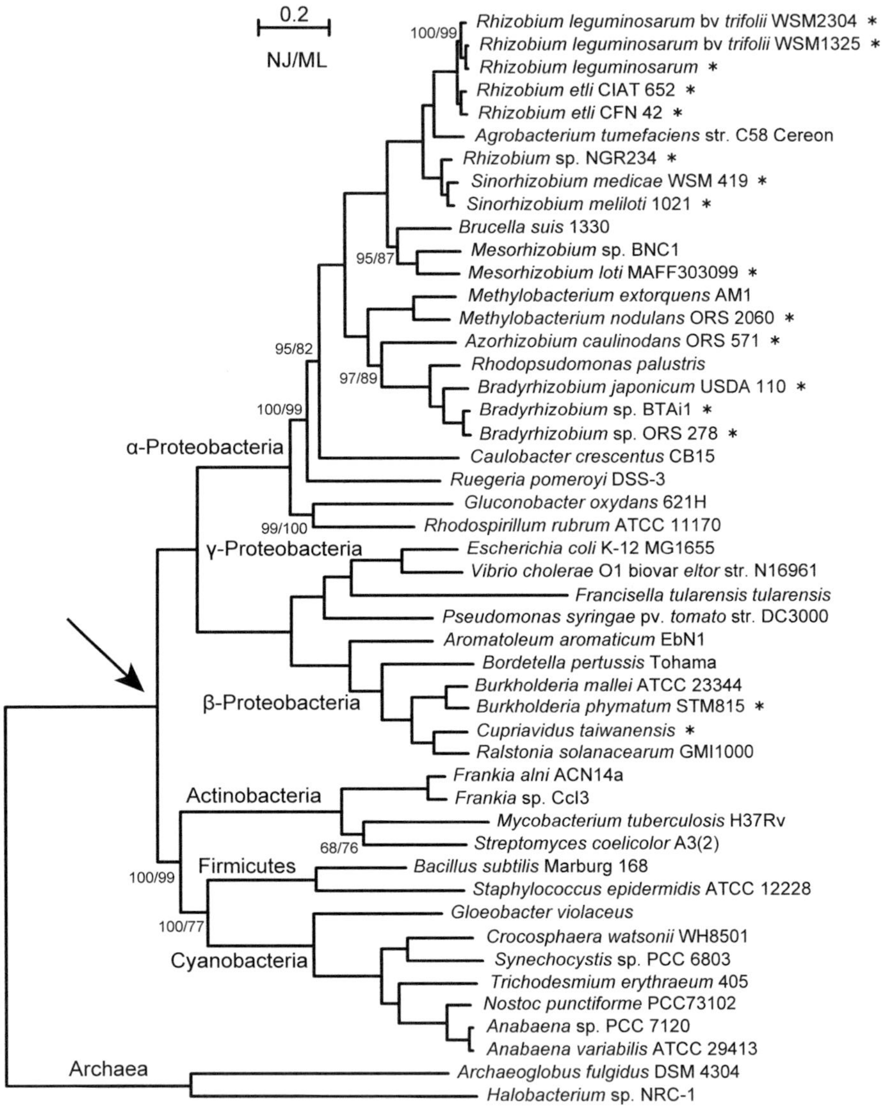

Fig. 6.19 Phylogenetic tree of prokaryotes. This phylogenetic tree was obtained by the Bayesian Inference method using concatenated 48 protein sequences. All branches are supported with 100% confidence level. This tree was originally made to analyze relationship of nodulating bacteria (asterisks). Green photosynthetic bacteria are not included. The position of the root of diversification of bacteria is shown by an arrow. Many nodulating bacteria harbor nodulation genes on indigenous plasmids, and therefore nodulation ability is found in phylogenetically unrelated strains in the α- and β-proteobacteria. The scale on the top indicates the evolutionary distance in which 0.2 amino acid substitutions occur on average. Phylogenetic trees were also constructed by the NJ and ML methods. The confidence values (in percentage) obtained with these methods were all 100/100 except otherwise indicated. Original figure

from which the evolution is supposed to start remains undetermined. Another type of information must be used to determine the root. In phylogenetic trees including only bacterial sequences, for example, it might be difficult to identify the root, and we have to be satisfied with unrooted trees showing relative relationship of taxa. Tree-drawing software has an option to select the position of root, and thus allows the users to show differently looking phylogenetic trees. Note, however, that the branching pattern itself does not change by the choice of root position. It is, therefore, important to examine the phylogenetic trees presented in research papers carefully by changing the root by imagination. Some tools are provided to estimate the root. MrBayes provides such an option. Tria et al. (2017) also developed an easy-to-use program to root a phylogenetic tree, by assuming that the root must be located at the center of a phylogenetic tree. The premise is valid if evolutionary rates do not vary extensively depending on branches.

Another simple method uses biological knowledge in estimating the root. In Fig. 6.19, the root must be located between the Archaea and the Bacteria, because they are fundamentally different in the origin. In other words, the root must be located somewhere on the vertical line (the length of this line has no meaning) or on the horizontal lines connected to it. Unfortunately, we have no means to identify the position of root exactly within these lines. Nevertheless, as long as the Archaea is the outgroup, the root of the bacterial clade (a clade is a group of taxa) is indicated by the arrow. An outgroup is a clade of organisms or proteins/genes that is clearly different from the rest of the organisms or proteins/genes (ingroup) under study. By selecting an outgroup, we can argue evolutionary history within the ingroup.

Thus, we can understand that bacterial evolution begins at the point indicated by the arrow, and two branches are separated (the upper and the lower clades). The upper clade includes the proteobacteria, whereas the lower clade consists of the actinobacteria, the firmicutes, and the cyanobacteria. The proteobacteria is further separated into the clades called α, β, γ, etc. It should be noted that a phylogenetic tree made with one family of proteins or genes represents the relationship of proteins or genes (protein or gene tree), but not of organisms. The tree in Fig. 6.19 was made with concatenated protein sequences to estimate the relationship of species (organism or species tree). But we have to be cautious. To estimate the phylogenetic relationship of organisms, we have to compare the trees of different proteins or genes, and to confirm that all of them give an identical branching pattern. As we will discuss later on the phylogenetic relationship between cyanobacteria and chloroplasts, different forms of phylogenetic trees might be constructed for different proteins or genes. Especially of note is a case of gene duplication in which two similar genes are produced at one point of evolution. Both of them might be kept in some clade, while one of them could be lost in another clade. Another one could be lost in the other clade. In this case, the gene tree does not match the organism tree. A well-known example is the D1 protein of Photosystem II (encoded by the *psbA* gene), which is present in multiple copies in various organisms (Qiu et al. 2012). In this case, we cannot deduce phylogenetic relationship of organisms from the phylogenetic tree of D1 proteins.

In examining phylogenetic trees, we also have to pay attention to the number shown at each branching point. It is expressed in real number (decimal) or in percentage. If the number is equal to 1.0 or 100%, the branching is highly reliable. As the number decreases, the reliability diminishes. In other words, the number indicates the reliability that the proteins/genes present in the branch form a single clade. The actual method of calculation of this reliability depends on the method of phylogenetic estimation. Bootstrap value is a classical reliability index used in the NJ method or parsimony method, indicating the proportion in which the particular branch is supported among many (usually 1000) calculations with resampling. A usual practice of phylogenetic analysis involves estimations with different methods. If only a single, most reliable tree is presented, several reliability indices obtained with different methods are presented at each branch. In many phylogenetic trees in this book, I present BI trees with reliability indices obtained by the BI and ML methods, as "value BI/value ML" except otherwise specified. In very crowded parts of the tree, the reliability indices are omitted for visibility. If the value is below 50%, the branch is not supported. Namely, depending on calculation methods, another member is included in the clade, or a member is excluded from the clade. We should be cautious about the reliability values in examining phylogenetic trees in research papers. Unfortunately, many experimental biologists do not understand this important caution. We sometimes find phylogenetic trees without reliability values in molecular biology articles. Such trees are arbitrary illustrations that do not represent phylogenetic relationship.

That is all for basic explanations on phylogenetic trees. Now we will see recent phylogenetic findings. Phylogenetic relationship of eukaryotic organisms has been studied extensively. Cavalier-Smith (2000) proposed a dichotomy Bikonta–Opisthokonta, which changed fundamentally eukaryotic taxonomy. More recently, a phylogenetic tree including hitherto unclassified organisms, such as Discicristata and Jakobids, was proposed (He et al. 2014, Fig. 6.20), and the root of all eukaryotes was estimated (between "a" and "b").

Then, how about chloroplasts? Figure 6.21 shows a Bayesian Inference (BI) tree using concatenated DNA sequences (excluding the third codon positions) encoding 16 proteins shared by bacteria, cyanobacteria, and chloroplast genomes. This can be considered a species tree. Placing various bacteria in the outgroup, we find a clade of cyanobacteria, in which chloroplasts diverge at the point indicated by the arrow. Green algae (chlorophytes), red algae (rhodophytes), and glaucophytes form a single clade. Similar trees have also been presented by various researchers using different data and methods. This type of tree has been taken as evidence for cyanobacterial origin of chloroplasts and monophyly (forming a single clade) of all chloroplasts.

Problems still remain. Exact form of the tree varies depending on various conditions, such as selection of genes, use of DNA or protein, or inclusion/exclusion of the third position of codons in DNA sequences. Figure 6.22 shows such examples. The upper tree was estimated with the 16S rRNA (one of the major RNA components in ribosome), whereas the lower one was with concatenated amino acid sequences of 11 proteins. Selection of organisms and selection of proteins/genes result in different forms of phylogenetic trees. In the lower tree in Fig. 6.22 using 11

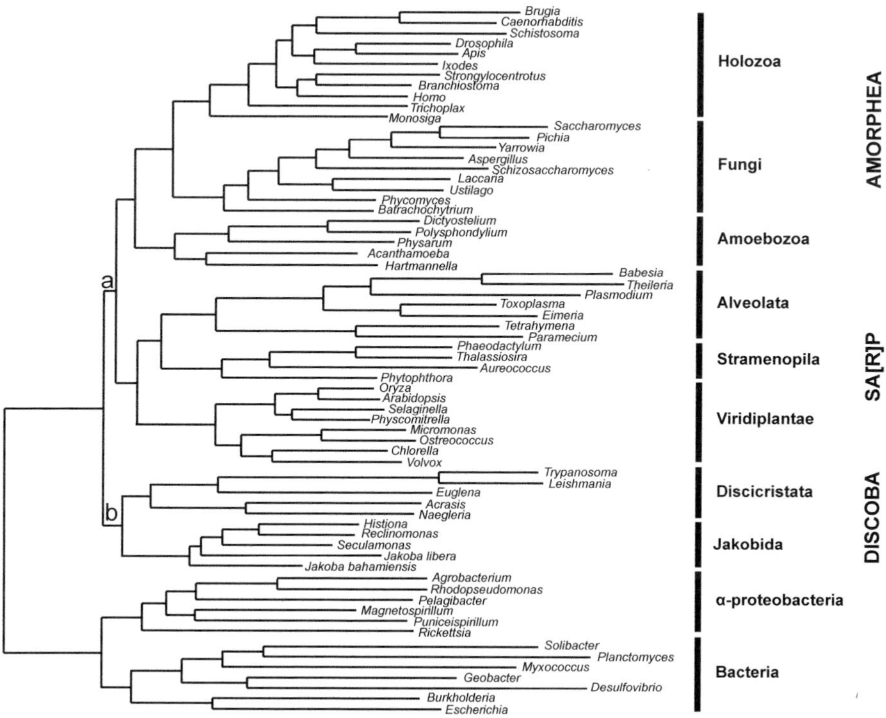

Fig. 6.20 A phylogenetic tree of all eukaryotes (He et al. 2014). Traditionally, eukaryotes have been classified into Bikonta (Alveolata, Stramenopila, Viridiplantae) and Opisthokonta (Holozoa, Fungi, Amoebozoa). In this tree, they form the clade (**a**). Another group of primitive microbes was identified (**b**), and the root of eukaryotes was placed between (**a**) and (**b**). Reproduced with permission of the publisher

protein sequences, *Cyanophora paradoxa* (a glaucophyte) is located within the clade of chloroplasts. Namely, chloroplasts are first separated into a clade of green algae and a clade consisting of red algae and *Cyanophora*. In contrast, in Fig. 6.21, *Cyanophora* diverges first, and then green algae and red algae are separated. The upper tree in Fig. 6.22 (ribosomal RNA) is still different from the two trees. Such inconsistency of trees as shown here is a serious problem that researchers are trying to resolve. Accordingly, some researchers consider red algae as the first diverging group, whereas the others believe that glaucophytes is the most basal group.

Nevertheless, one thing is common to all these trees: namely, green algae (including green plants), red algae, and glaucophytes are monophyletic (i.e., they belong to a single clade). In other words, these three lineages of photosynthetic organisms diverged from a common ancestor. Although only data with chloroplast-encoded genes are presented here, researchers also studied phylogenetic analysis based on nuclear-encoded protein sequences, supporting monophyly of the three groups of organisms (Moreira et al. 2000; Lopez-Garia et al. 2017). Accordingly, these three groups of organisms are considered descendants of an organism born as a result of primary endosymbiosis, and called Archaeplastida.

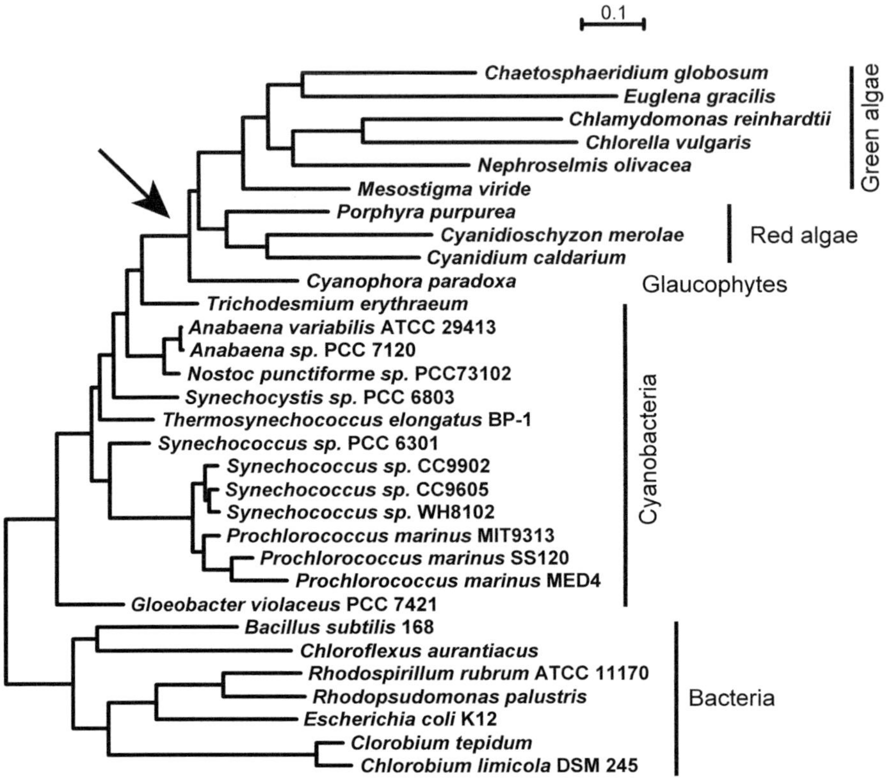

Fig. 6.21 Phylogenetic tree of concatenated DNA sequences encoding 16 proteins shared by chloroplasts, bacteria, and cyanobacteria. This tree was calculated by the BI method using the first and the second positions of codons of the coding sequences (nucleic acids). All branches are supported at 100% confidence. The arrow corresponds to the birth of chloroplasts. Original figure in Japanese version

Another remaining problem concerns the nature of the cyanobacterium that became the endosymbiont in the primary endosymbiosis, namely, the ancestor of all chloroplasts. The figures presented above showed that the first chloroplast diverged from a cyanobacterium similar to *Anabaena* sp. PCC 7120, filamentous species (many cells are arranged like a filament) that fixes atmospheric nitrogen. A recent study using many cyanobacterial species including primitive cyanobacteria indicated that chloroplasts diverged soon after the diversification of basal clades (Shih et al. 2013). Figure 6.23 shows diagrammatically the primary and secondary endosymbioses according to new data. A more recent study proposed that a freshwater strain *Gloeomargarita lithophora* is the most related to the ancestor of chloroplasts (Ponse-Toledo et al. 2017). There is still disagreement among the researchers on the exact cyanobacterial species that is closest to the origin of chloroplasts. A new candidate or candidates might be proposed in the near future.

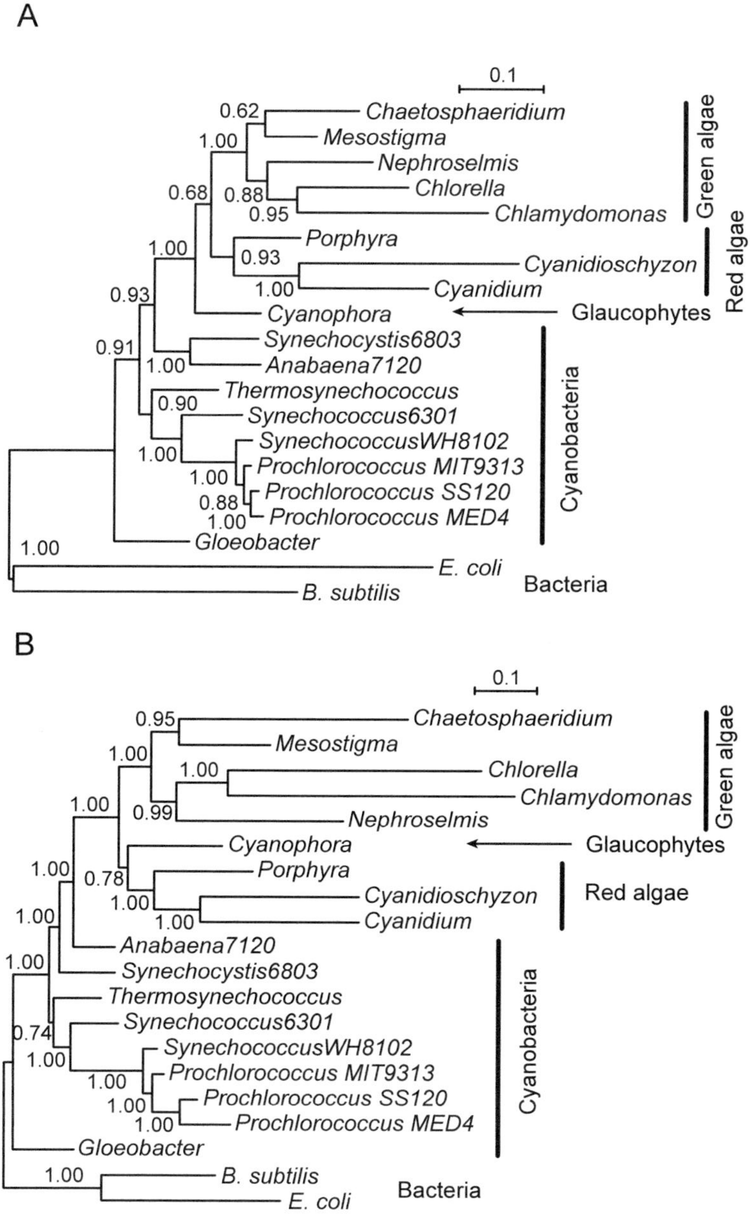

Fig. 6.22 Phylogenetic trees of 16S rRNA (**a**) and concatenated 11 proteins (**b**) using 20 taxa. Both trees represent results of the BI method. Scale indicates evolutionary distance. Confidence values of branches are shown in decimal values. In the rRNA tree, *Cyanophora* is the first taxa that diverged in the chloroplasts. In the protein tree, there are two clades: the one consisting of red algae and glaucophytes, and the other consisting of green algae. Original figure in Japanese version

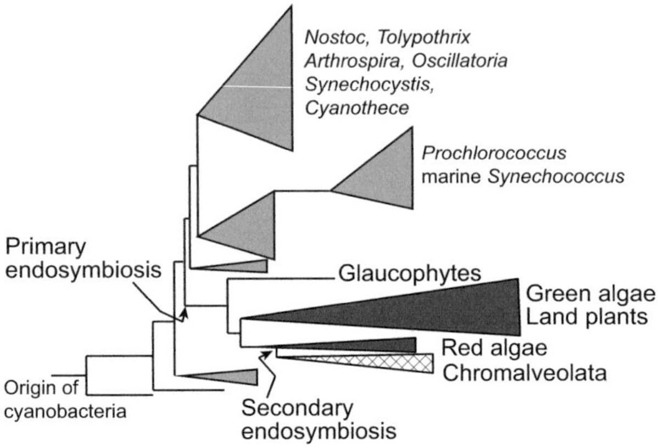

Fig. 6.23 Diagram of phylogenetic relationship between cyanobacteria and plastids based on new data. Original figure based on Shih et al. (2013)

6.4 Resolution of the Two Problems Inherent in the Original Theory of Margulis

Progress in molecular phylogenetic analysis resolved the two problems inherent in the original theory of Lynn Margulis in 1970 (see Chap. 4). One of them is related to the "botanical myth" or refutation of the conventional plant taxonomy, and the other on the multiple parallel endosymbioses involving many different chloroplast endosymbiotic events.

6.4.1 The Truth in "Botanical Myth"

As described earlier in Sect. 4.1, Margulis (1970) denied the conventional view of plant taxonomy/phylogenetics on the evolutionary path from cyanobacteria, through algae, until land plants, and stated that algae and plants were formed by endosymbiosis of chloroplasts after diversification of eukaryotes. In this case, she believed simply that introduction of chloroplasts into animal cells created plant cells, because both animals and plants were included in the major lineage of eukaryotes. However, with the advances in molecular phylogenetics, Cavalier-Smith (2000) proposed the distinction between the Bikonta and the Opisthokonta, thus splitting the entire world of eukaryotes into a supergroup consisting of algae and plants and the other consisting of fungi and animals. The Bikonta is a collection of organisms having two anterior flagella, whereas the Opisthokonta includes organisms having a single posterior flagellum. Some organisms do not have flagella at all, but the molecular phylogenetic relationship with flagella-bearing organisms helped attribution of most organisms to one of the two supergroups. As Fig. 6.20 indicates, we are now familiar with a more detailed classification of all eukaryotes. Curiously, almost all chloroplast-harboring organisms (except *Euglena*) are included in particular groups of organisms, namely, Alveolata, Stramenopiles, and Plants plus Algae (Archaeplastida). Endosymbiosis of chloroplasts is limited to these groups of organisms (formerly,

Bikonta). Based on this observation, Professor Hisayoshi Nozaki (University of Tokyo), who collaborated with us on the genome sequencing project of *Cyanidioschyzon merolae*, proposed the notion Plantae, or an extended notion of plants. No chloroplasts are found in Amorphea including animals and fungi.

This reflection clearly indicates that there are different kinds of eukaryotes, including those which are able to become photosynthetic and those which are not. The "botanical myth" was not entirely erroneous, but kept some truth. As described in the next section, the algae having the primary endosymbiont and those having the secondary one originated from different hosts, but both of these hosts belonged to the Bikonta or SA[R]P (+Discicristata), and were related in a broad sense. This was a consequence of molecular phylogenetics that Margulis did not anticipate.

6.4.2 Secondary Endosymbiosis Explaining Multiple Parallel Endosymbiosis

In the original hypothesis of Margulis, chloroplast endosymbiosis occurred many times in parallel in different lineages of eukaryotes. Martin (2017) counted as many as 20 independent endosymbiotic events in the phylogenetic tree of Margulis. Curiously, she rejected the taxonomy of Copeland in her 1970 book (p. 274 in Chapter 9, Margulis 1970) dividing green algae and land plants according to the mode of mitosis (he placed green algae in protists), and considered they arose by a single endosymbiosis by a "green" blue-green algal cell. Currently, however, researchers consider that a single endosymbiosis engendered the Archaeplastida. Namely, land plants, green algae, red algae, and glaucophytes all have chloroplasts originating from an ancestral chloroplast generated by a single endosymbiotic event capturing a cyanobacterium (the primary endosymbiosis). In contrast, brown algae and diatoms (heterokonts) are thought to be descendents of a product of endosymbiosis of red algal cell within a eukaryotic cell, which left only the red algal plastids in the cytoplasm. This is called the secondary endosymbiosis (Fig. 6.23, hatched clade). A putative intermediate stage is found in cryptophytes (Fig. 6.24), in which red algal nucleus remains as a nucleomorph. The notion of secondary endosymbiosis was indeed proposed to explain the cell architecture of cryptophytes (Whatley 1981; Cavalier-Smith 1982). Another type of secondary endosymbiosis includes a green algal cell: *Euglena* was produced by endosymbiosis of a green alga within a non-photosynthetic cell (Discicristata); Chlorarachniophytes are also putative products of green algal endosymbiosis within a Cercozoa cell, keeping both chloroplasts and nucleomorph. Furthermore, tertiary endosymbiosis has been hypothesized for the origin of haptophytes and dinoflagellates. The origin of all these diverse algae is now explained by the notion "secondary endosymbiosis." Interestingly, the apicoplast within the malaria parasite *Plasmodium falciparum* is also supposed to be a plastid originating from the secondary endosymbiosis (for a review, see McFadden and Gilson 1995). A new system of algal taxonomy was established on the basis of the primary and secondary endosymbioses (Adl et al. 2005) and revised recently (Adl et al. 2019).

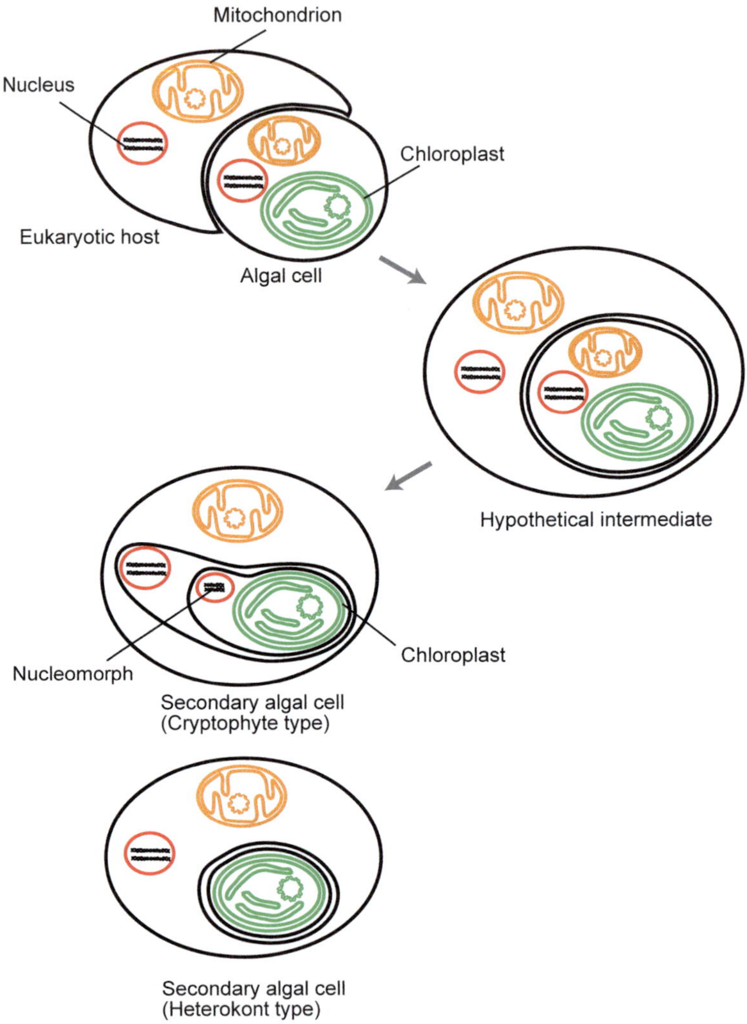

Fig. 6.24 The secondary endosymbiosis explains the origin of cryptophytes and heterokonts. Similar diagrams were presented in Cavalier-Smith (1982) and Douglas et al. (2001)

Cavalier-Smith (2000) recognized the importance of membrane heredity in interpreting endosymbiosis. He presented clearly color-coded figures to show that the endosymbiont membranes were kept intact during the entire evolutionary history that followed until now. Such attitude to emphasize the membrane correspondence is widely found in many textbooks. That is why students misunderstand that chloroplast membranes are identical to cyanobacterial membranes, or even that chloroplasts are cyanobacterial cells living within a plant cell.

6.4.3 Discovery of Multiple Independent Primary Endosymbioses

As explained earlier, most part of the multiple parallel endosymbioses that Margulis conceived was not supported by current data, but other instances of primary endosymbiosis became evident. They are chromatophores of some rhizopods and spheroid bodies of some diatoms (both were described in Mereschkowsky (1905a). See Sect. 2.4 for the history). Margulis (1970) did not mention them clearly in her very large phylogenetic tree (Fig. 4.15). Most probably, she just tried to explain the origin of differently colored plastids by endosymbiosis of differently colored cyanobacteria, without considering chromatophores and spheroid bodies.

The pair of sausage-shaped chromatophores in the rhizopod cell of *Paulinella chromatophora* is now believed to originate from a recent primary endosymbiosis (about 60-100 Mya) (Fig. 6.25). The cyanobacterial endosymbiont is considered to be close to the α-cyanobacteria that are major marine producers. Therefore, the origin of the chromatophores of *Paulinella* is definitely different from the origin of Archaeplastida plastids (Fig. 6.23). Currently, the chromatophore genomes of two strains (*P. chromatophora* and *Paulinella* sp. DL 2016a; Nowack et al. 2008; Lhee et al. 2017, respectively) and the nuclear genome of *P. chromatophora* (Nowack et al. 2016) have been reported. Phylogenetic trees as shown later clearly indicate that *Paulinella* chromatophores diverged from cyanobacteria, and are very similar to cyanobacterial cells virtually living inside the cells.

Another case of primary endosymbiosis was discovered in some diatoms, in which colorless cyanobacterial remnant called spheroid body exists within the cytosol and plays a role in nitrogen fixation. A similar, but clearly distinct case is the mysterious cyanobacterium called UCYN-A, which was discovered by metagenomic analysis of marine water (Zehr et al. 2008). The genome was reconstructed by computer, but the organism remained unknown. Nitrogen-fixing granules in a haptophyte was found to be responsible for the genome UCYN-A (see Introduction, Table 1.2: Hagino et al. 2013). Currently, many works are performed on various types of curious organisms that harbor cyanobacteria-related endosymbionts.

Interestingly, the discovery of chromatophores and spheroid bodies dates back to the early twentieth century. Mereschkowsky (1905a) described them as examples of cyanobacterial endosymbionts (Sect. 2.4). He also considered the blue-green cyanelles of a cryptophyte *Cyanomonas* cyanobacterial endosymbionts. A glaucophyte *Cyanophora paradoxa* was discovered later (Korschikov 1924) (Fig. 6.26). *Cyanophora* and *Paulinella* were known as representative algae that harbored cyanobacterial endosymbionts called cyanelles until the 1970s (Stanier et al. 1957/1976). Cyanelles are surrounded by a peptidoglycan layer, like cyanobacteria. The cyanelle of *Cyanophora* turned out to be a common chloroplast by the analysis of cyanelle genome. *Paulinella* is, however, a product of a recent endosymbiosis as explained above. The chromatophore genome is as large as 1 Mbp, about a half of the smallest cyanobacterial genome. Members of *Paulinella* are ubiquitous organisms that are also found in rice fields of Japan.

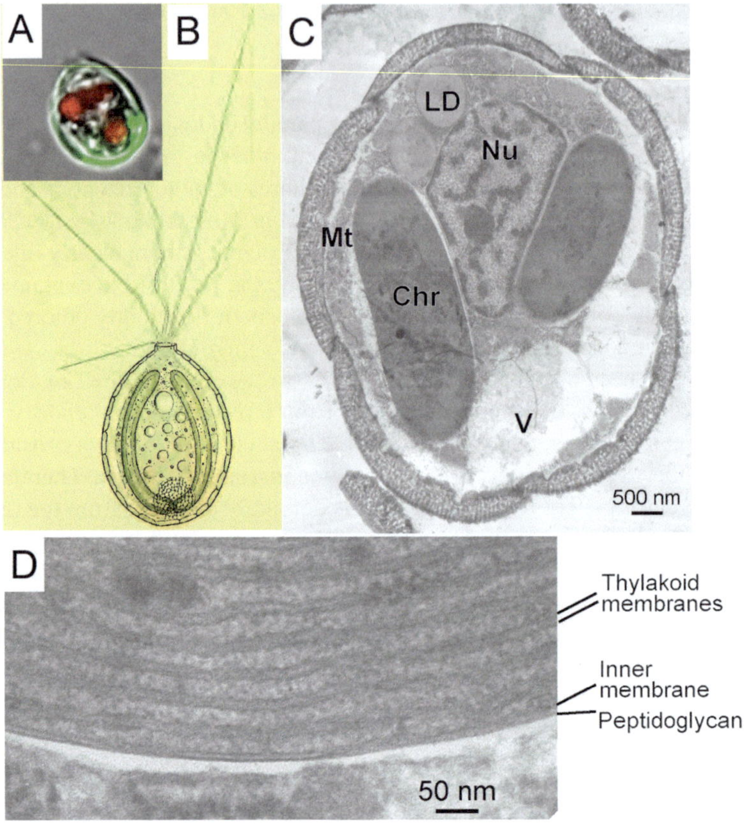

Fig. 6.25 Images of *Paulinella micropora* MYN1 and historical drawing of *Paulinella chromatophora* (Lauterborn 1895). (**a**, **c**, and **d**) *Paulinella micropora*. (**b**) *Paulinella chromatophora*. Panel (**a**) is a merged image of fluorescence (green, silicate; red, chlorophyll) and differential interference (taken by Masashi Yamazaki in the laboratory). Panel (**b**) is an original drawing by Lauterborn (1895). Panels (**c** and **d**) are transmission electron micrographs (taken by myself). Panel (**d**) is an enlarged view of peripheral part of a chromatophore. Lauterborn (1895) and Pascher (1929a) described isolation of chromatophores by cell breakage, suggesting that the chromatophores were stable without osmoticum that we normally use in cell fractionation. Chr, chromatophore; LD, lipid droplet; Mt., mitochondrion; Nu, nucleus; V, vacuole

There are abundant examples of secondary endosymbiosis and similar phenomena. The green *Paramecia* and the corals presented in Chap. 1 are examples of reversible endosymbiosis of *Chlorella* (green alga) and *Symbiodinium*. They are similar to secondary endosymbiosis because a eukaryotic cell harbors another species of eukaryotic cells. *Hatena arenicola* is another strange example of reversible secondary endosymbiosis (Fig. 6.27: Okamoto and Inoue 2006; Yamaguchi et al. 2014).

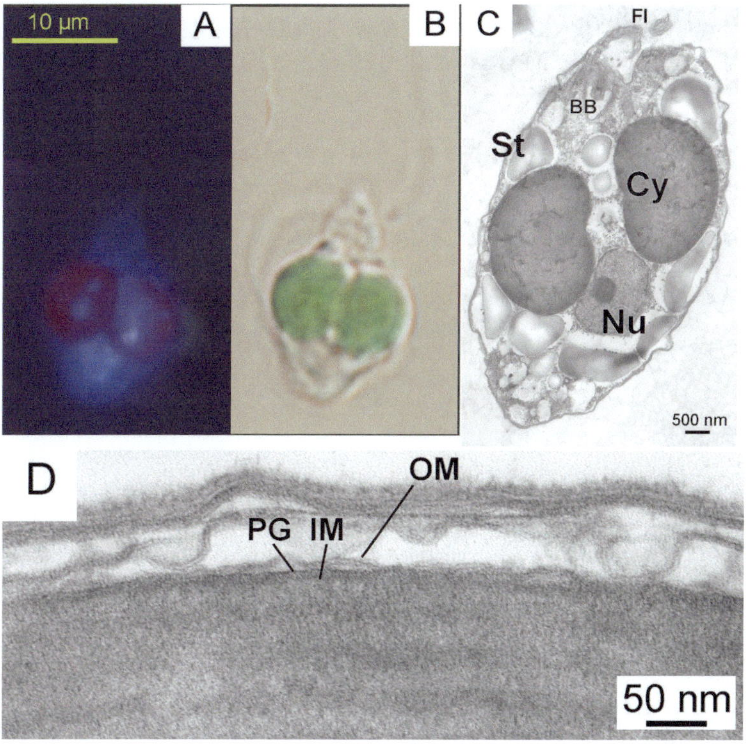

Fig. 6.26 Images of *Cyanophora paradoxa* NIES-547. Panels (**a** and **b**), fluorescence and differential interference images, respectively. Two large cyanelles are visible. Blue-white fluorescence, DNA; red fluorescence, chlorophyll and phycocyanin. Panels (**c** and **d**), transmission electron micrographs. BB, basal body; Cy, cyanelle; Fl, flagellum; Nu, nucleus; St, starch granule. IM, inner envelope membrane; OM, outer envelope membrane; PG, peptidoglycan layer. Original figure

Note that "Hatena" is a Japanese word meaning "mysterious." In this case, a prasinophyte *Nephroselmis rotunda* is incorporated into a marine flagellate *Katablepharis*. Since the partially digested endosymbiont does not divide within the host, only one of the daughter cell retains the endosymbiont after cell division. The other daughter cell will acquire a new endosymbiont. This was a discovery of the group of Professor Isao Inoue (Tsukuba University) in Isono-ura Bay (Wakayama Prefecture, Japan). Details were given in Japanese in Inoue (2006/2007J, pp. 323–327).

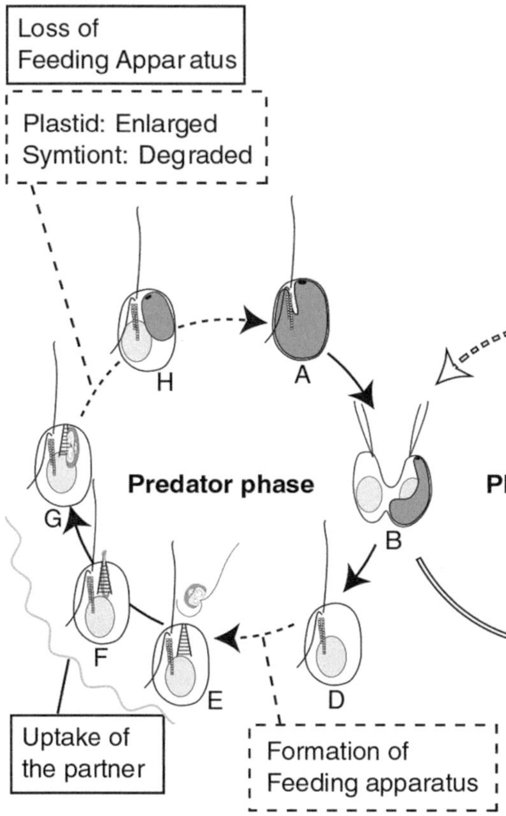

Fig. 6.27 Life cycle of *Hatena arenicola*. Uncolored cell (**d**) ingests a *Nephroselmis* cell using a special feeding apparatus consisting of microtubules. The symbiont is partially digested to give nucleus, mitochondrion, and chloroplast. Then the chloroplast is enlarged and becomes active in photosynthesis (**a**). Upon cell division (**b**), one of the daughter cells receives the chloroplast (**c**), whereas another becomes colorless (**d**). Reproduced from Okamoto and Inoue (2006) with permission of the publisher

6.5 Origin of Mitochondria

Readers might question why mitochondria are not the main subject in this book on the endosymbiotic theory. The reason is simple: cyanobacterial origin of chloroplasts seems far easier to understand and to support, because photosynthetic mechanism is certainly common in cyanobacteria and chloroplasts. Similarity of mitochondria and bacteria might not be evident. Even the mechanism of respiratory chain is not exactly the same. Support for the bacterial origin of mitochondria is very poor. It is difficult to determine when and by whom endosymbiotic origin of mitochondria was established. If endosymbiotic theory of chloroplast origin is denied or modified (see next chapter), then we will have less solid support for the mitochondrial origin. However, in this section added in English version, I briefly explain the situation in the search for the origin of mitochondria.

6.5.1 Brief History

As explained in Part I (Sect. 3.3), endosymbiotic origin of mitochondria was nothing more than a hypothetical idea without solid evidence in the first half of the twentieth century. The nature of mitochondrion remained unclear until the elucidation of its ultrastructure by electron microscopy and the biochemical characterization of isolated mitochondria in the 1950s and 1960s (see Stanier and van Niel 1962 for a review). The demonstration of respiratory activity of mitochondria with ATP synthesis was an important step towards the recognition that a mitochondrion is an independent metabolic unit (Sect. 5.2). Then, the discovery of mitochondrial DNA reminded of an ancient idea that mitochondria were symbiotic bacteria. Schwartz and Dayhoff (1978) gave the first convincing phylogenetic tree with 5S rRNA and cytochrome *c* that supported bacterial similarity of mitochondria. Nevertheless, monophyletic relationship of all mitochondria remained to be established.

As explained already, Gray (1992) affirmed that all mitochondria were monophyletic and that the ancestor of all mitochondria was related to α-proteobacteria, based on the phylogenetic tree of small subunit (SSU) rRNA. Note that mitochondrial ribosomal RNA is variable in size: rRNA of the SSU rRNA (homolog of bacterial 16S rRNA) ranges from 12S (mammals) to 18S (plants). Human mitochondrial genome encodes only 13 proteins. Plant and algal mitochondrial genomes encode some more proteins. An early phylogenetic analysis using mitochondrial cytochrome *b* and cytochrome *c* oxidase subunit I suggested a mitochondrial origin within Rickettsiaceae (Sicheritz-Pontén et al. 1998). Phylogenetic analysis using 15 mitochondria-encoded protein sequences of three selected organisms showed a meaningful grouping of mitochondria with Rickettsiales (Fitzpatrick et al. 2006). Phylogenetic analysis of 29 proteins shared by algal mitochondrial genomes suggested a single origin of these mitochondria and their affiliation to α-proteobacteria, but not specifically to *Rickettsia* (Tajima et al. 2016). A recent analysis using an extended set of α-proteobacterial sequences (including new metagenome data) resulted in a deep origin of mitochondria outside the known α-proteobacterial lineages (Martijn et al. 2018). The origin of the common ancestor of all mitochondria remains to be identified.

6.5.2 Proteome

Phylogenetic analysis of the whole mitochondrial proteome, however, presented a different perspective. Although the mitochondrial genome is small with its coding capacity of less than 100 proteins (72 in *Andalucia godoyi* and 67 in *Reclinomonas americana*: Burger et al. 2013), about 1000 proteins are supposed to exist and function in mitochondria (Gray 2015). In an early analysis of yeast mitochondrial proteome (423 proteins were identified at the time), about a half of them were found to be of the eukaryotic origin (Karlberg et al. 2000). Gray (2015) showed that only a small proportion (10–20%) of the mitochondrial proteome displayed a clear α-proteobacterial ancestry. Pittis and Gabaldón (2016) showed that, among the 1078 superfamily of proteins shared by the last eukaryotic common ancestor (LECA), only 79 were estimated to originate from α-proteobacteria. According to the statistics of the stem length (corresponding to the time elapsed from the acquisition of the protein family to the beginning of diversification of eukaryotes), the mitochondrial proteins of α-proteobacterial ancestry showed the shortest phylogenetic distances to the proteins of other bacterial origins (except those of cyanobacterial origin), suggesting a later acquisition of mitochondria by a host that already had a mitochondrial proteome acquired from various bacteria. Harish and Kurland (2017) constructed three phylogenetic trees of concatenated protein families from about 50 genomes, one with the 614 protein families having homologs in mitochondria, another with the 902 protein families having no mitochondrial homologs, and the other including all protein families. The three trees were virtually identical, indicating that both mitochondrial and non-mitochondrial proteins had an identical evolutionary history. The authors concluded that mitochondria are not captive bacteria. Currently, various versions of hypotheses are presented to explain chimeric nature of mitochondrial proteome (Lang and Burger 2012; Gray 2014; Roger et al. 2017).

6.5.3 Mitochondrial-Remnant Organelles

Another complication is the diversity of mitochondria-related organelles. Hydrogenosomes that produce hydrogen were recognized as anaerobic homologs of mitochondria, which are aerobic organelles (see Tielens et al. 2002; Martin and Müller 2007 for early reviews). Hydrogenosomes are small organelles (200 nm to 1 μm in diameter) found in *Parabasalia* protozoa and some other protists (Benchimol 2009). They do not contain DNA, but are surrounded by two membranes that contain cardiolipin, like mitochondria. Other related organelles are also known in phylogenetically unrelated eukaryotes (Roger et al. 2017), such as anaerobic mitochondria, hydrogen-producing mitochondria, and mitosomes devoid of DNA. Activity of FeS cluster assembly is mostly conserved in all these different types of mitochondrial-remnant organelles (MRO). The origin of MRO without DNA cannot be determined by rRNA sequences, but phylogenetic analysis using some conserved proteins showed that hydrogenosomes and mitosomes are indeed related to mitochondria (see Chapters 6 and 7 of Martin and Müller 2007).

6.5.4 Lipid Biosynthesis

Next, I explain the origin of mitochondrial membrane lipids (Harvath and Daum 2013). Mitochondria contain glycerophospholipids, such as phosphatidylcholine (PC), phosphatidylethanolamine (PE), phosphatidylglycerol (PG), and cardiolipin (CL). In plants, animals, and fungi, PC and PE are synthesized in the extramitochondrial compartment except for phosphatidylserine decarboxylation (yielding PE) and cytidylyltransfer to phosphoethanolamine. CL, which is specific to mitochondria and MRO (except *Giardia* mitosomes), is synthesized via PG. PG synthesis is performed in mitochondria and endoplasmic reticulum, as well as chloroplasts in plants and algae. Bacteria (α-proteobacteria) contain PE, PG, and CL as a standard set, but additional lipids (PC and glycolipids) are also present in some species. The pathway of PG and CL synthesis seems common in bacteria and mitochondria, but phylogenetic analysis indicates that bacteria and mitochondria contain different types of CL synthases (CLS): Eukaryotic enzymes are homologs of prokaryotic PG synthases, whereas bacterial CL synthases are entirely different enzymes (Sato 2019). Curiously, putative CLS of some α-proteobacteria is closely related to eukaryotic CLS, but many α-proteobacteria including *Rickettsia* do not possess eukaryotic CLS homolog. In the *Rickettsia* metabolic profiles, CLS remains blank, and the authors estimated that CL might be supplied by the host (Driscoll et al. 2017). Mechanism of PE synthesis is also different in bacteria (phosphatidylserine decarboxylation) and plant/algal mitochondria (phosphoethanolamine pathway). Bacterial PC synthesis involves phosphatidylcholine synthase, whereas eukaryotic PC synthesis takes place either by PE methylation or via the pathway involving phosphocholine (Sato et al. 2016; Hirashima et al. 2018). These differences in biosynthetic mechanisms (Fig. 6.28) suggest that mitochondrial lipid synthesis was not inherited from bacterial endosymbiont (if endosymbiosis did happen). Rather, eukaryotic host already possessed lipid biosynthesis system for mitochondria before acquisition of mitochondria. This is clearly understood in the case of hydrogenosomes, in which no organellar DNA is present. This is the same situation in chloroplast membrane lipids as I will explain in detail in the next chapter. Therefore, mitochondrial membrane lipids are not products of prokaryotic genes, and hence the mitochondrial membranes as clearly shown in diagrams of endosymbiosis are not membranes of endosymbionts. It is important to recognize that "membranes are not inherited by themselves." Accordingly, a recently proposed bacterial vesicle secretion hypothesis that mitochondrial endosymbiont generated internal membranes of eukaryotes (Gould et al. 2016) is not supported. Mitochondrial lipids are not synthesized by bacterial systems.

6.5.5 Conclusion

In summary, the origin of mitochondria must be far more complex than it was thought at the end of the twentieth century. We should consider many prokaryotic and eukaryotic genes that had been present in the eukaryotic host before the formation of mitochondria. We should also take into account many gene transfers and gene losses after the establishment of mitochondria. We still have to study the mitochondrial ancestor, which might not be similar to *Rickettsia* that was easily hypothesized as a candidate because of its parasitic property. In the next chapter, we will examine "the cyanobacterial ancestor of chloroplasts," which is supposed to be more plausible than bacterial ancestor of mitochondria.

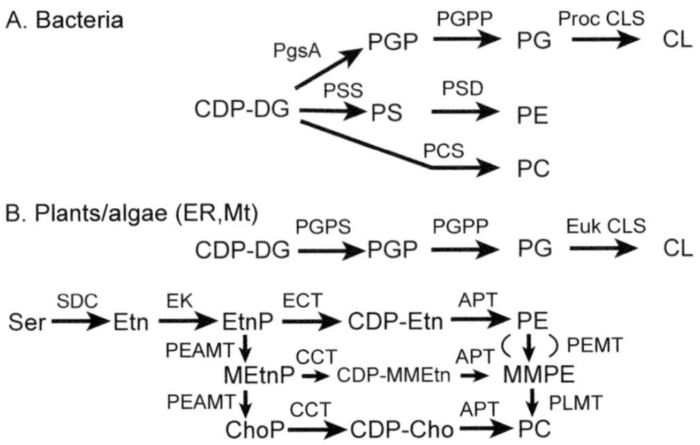

Fig. 6.28 Pathways of phospholipid biosynthesis in bacteria and plants/algae. Phospholipid biosynthesis is essentially different in bacteria and plants/algae, except PG synthesis. PC is not present in all bacteria. PEMT is not known in *Arabidopsis* and other plants, but present in *Cyanidioschyzon merolae* (red alga). Some green algae use PLMT for this reaction (Hirashima et al. 2018). Substance names: CDP-Cho, cytidinediphosphocholine; CDP-DG, cytidinediphosphodiacylglycerol; CDP-Etn, cytidinediphosphoethanolamine; CDP-MMEtn, cytidinediphosphomonomethylethanolamine; ChoP, choline phosphate; CL, cardiolipin; Etn, ethanolamine; EtnP, ethanolamine phosphate; MEtnP, monomethylethanolamine phosphate; MMPE, phosphatidylmonomethylethanolamine; PC, phosphatidylcholine; PE, phosphatidylethanolemine; PG, phosphatidylglycerol; PGP, phosphatidylglycerol phosphate; PS, phosphatidylserine. Enzyme names: APT, aminoalcohol phosphotransferase; CCT, choline phosphate cytidylyltransferase; CLS, cardiolipin synthase; ECT, ethanolamine phosphate cytidylyltransferase; EK, ethanolamine kinase; PEMT, phosphatidylethanolamine methyltransferase; PLMT, phospholipid methyltransferase; PCS, phosphatidylcholine synthase; PGPP, phosphatidylglycerolphosphate phosphatase; PGPS (PgsA), phosphatidylglycerolphosphate synthase; PSD, phosphatidylserine decarboxylase; PSS, phosphatidylserine synthase; SDC, serine decarboxylase. Mammals have another pathway of PE synthesis, namely, decarboxylation of PS produced by base exchange of phospholipids with serine

Chapter 7
Continuity and Discontinuity of Chloroplasts and Cyanobacteria

Genome sequencing opened a new way of search for chloroplast origin. The term "genome" originally meant the whole set of genes of an organism, but now it means the entire DNA of an organism. Most eukaryotes are diploid, and have two sets of genome. In organisms such as human having sex chromosomes, a genome consists of a set of autosomes (22 in human) plus X and Y sex chromosomes. In the 2000s, genome sequences of various model organisms such as human, *Arabidopsis thaliana* (thale cress), *Escherichia coli*, and budding yeast were determined. Then, new generation sequencers drastically cut the cost of sequencing and allowed sequencing of genomes in non-model organisms, namely, actual organisms studied in biology laboratories, at a reasonable cost. We are now in the post-genomic era. Additionally, various large-scale data, such as transcriptomes (data of all transcripts), proteomes (data of all proteins), and metabolomes (data of all metabolites), became available at least in model organisms. These advances enabled studies on biologically interesting phenomena by wisely combining all these "omics" data, not only in model organisms but also in non-model organisms. Comparative genomics is a new technique that the progress in computer technology realized. This technique enables to establish relationship between genomic data of different organisms, and to apply knowledge obtained in an organism to other organisms. It reveals many homologous proteins in all sequenced organisms, for use in phylogenetic analysis. In this way, post-genomic studies provided enormous new data for the study on the origin of organelles, which could, perhaps, change fundamentally the endosymbiotic theory on the origin of chloroplasts and mitochondria. This is a benefit of the post-genomic re-analysis over the studies performed in the latter half of the twentieth century.

As I stated repeatedly (Sato 2001, 2006), similarities and differences are found between cyanobacteria and plastids. Similarities should not be limited to apparent resemblances, but must imply revelation of the fundamental continuity of evolutionary history from cyanobacteria to plastids. In contrast, differences represent historical discontinuity between cyanobacteria and plastids (Chap. 6, Table 6.8). However, we have to understand another type of differences that I explain in detail here. Both Mereschkowsky and Margulis pointed out similarities between cyanobacteria and

N. Sato, *Endosymbiotic Theories of Organelles Revisited*,
https://doi.org/10.1007/978-981-15-1161-5_7

chloroplasts, which they regarded as a sign of continuity that supported endosymbiotic hypothesis of chloroplast origin. In the time of Margulis and in later times, similarities in biochemical constituents and protein synthesis system were also added to the supporting evidence for the hypothesis: namely, oxygenic photosynthesis (and the presence of chlorophyll), the presence of genetic material, the presence of protein synthesis system, as well as various prokaryotic characteristics of plastids (absence of nucleus, sensitivity to inhibitors of prokaryotic gene expression system, etc.). In fact, evaluation of these properties is not easy, because apparent similarity might not represent phylogenetic continuity. It is easy to understand that differences in properties and constituents are direct evidence against phylogenetic continuity. In addition, we have to understand that phylogenetic analysis of apparently similar characters, enzymes, and genes could be results of independent, parallel or convergent evolution, in other words, that they represent hidden discontinuity in the history of cyanobacteria and plastids. That is why we have to examine closely the characters of cyanobacteria and plastids, by focusing on the underlying mechanisms. I will show various types of examples in the following sections.

7.1 Glycolipid Biosynthesis System

In the interior of a chloroplast, thylakoid membranes in the form of flattened closed sacs are the sites of photochemical reactions. Various protein complexes are embedded in the membranes. Thylakoid membranes are also present in cyanobacterial cells. They are curved in concentric form (Introduction, Fig. 1.8b; Chap. 6, Fig. 6.25 shows thylakoid membranes of *Paulinella*). In plant chloroplasts, thylakoid membranes are stacked to form grana (Introduction, Fig. 1.6). The thylakoid membranes consist of lipid components and proteins. The latter includes PS I complex (PS stands for photosystem), PS II complex, cytochrome b_6f complex, and ATP synthase complex. Lipid components are glycolipids and phospholipids. Major glycolipids of thylakoid membranes are monogalactosyl diacylglycerol (MGDG), digalactosyl diacylglycerol (DGDG), and sulfoquinovosyl diacylglycerol (SQDG). SQDG is a curious sulfonolipid, containing S-C linkage, which is rare in nature. The major phospholipid component of thylakoid membranes is phosphatidylglycerol (PG). Phosphatidylethanolamine (PE) and phosphatidylcholine (PC), typical lipid components of eukaryotic cell, are not present (or very minor components) in the thylakoid membranes of plant chloroplasts. PC is known to exist at a low level in the envelope membranes, but PE is considered a cytoplasmic contaminant, if it is detected in chloroplast preparations. Glycolipids are also present in the photosystem complexes, and a small number of glycolipid molecules are considered essential in the photochemical reactions (for general description on chloroplast membranes and photosynthetic mechanisms, see textbooks on plant physiology, Buchanan et al. 2002; Taiz and Zeiger 2010, or textbook on photosynthesis, such as Hall and Rao 1972/1999; *The Science of Photosynthesis* 2007J; Sato 2014J).

Under these circumstances, the presence of common glycolipids (MGDG, DGDG, and SQDG) in chloroplasts and cyanobacteria was once considered an important piece of evidence for the cyanobacterial origin of chloroplasts. Carr and Craig (1970) explained similarities of lipid composition in chloroplasts and cyanobacteria, citing Sagan (1967). Nichols (1970), however, just discussed the function of lipids, in the framework of classical phylogenetic relationship of photosynthetic organisms (called "botanical myth" by Margulis), without mentioning endosymbiotic origin of chloroplasts.

Feige et al. (1980) in the group of Ernst Heinz (Germany) reported a glucolipid intermediate of MGDG synthesis. I also started to study lipid biosynthesis in cyanobacteria in 1978 as a graduate student, and proposed that cyanobacteria and chloroplasts synthesize galactolipids by different mechanisms (Sato and Murata 1982). All the enzymes involved in the biosynthetic steps have been identified now as shown in Fig. 7.29, which reveals fundamental differences in the mechanism of MGDG synthesis from diacylglycerol (DAG) in chloroplasts and cyanobacteria. In chloroplasts of plants and algae, MGDG is synthesized by direct galactosylation of DAG, whereas in cyanobacteria, DAG is glucosylated to form monoglucosyl diacylglycerol (GlcDG), which is isomerized (epimerized) to form MGDG. Namely, the glucosyl moiety is isomerized to galactosyl by changing the orientation of the OH group at the C-4, without cleaving the sugar-to-glycerol linkage. Phylogenetic analysis showed that the plant-type galactosyltransferases, MGD1 and DGD1, did not originate from cyanobacteria (see Sato 2015, 2019; Sato and Awai 2016). MGD1 originated probably from green photosynthetic bacteria (or simply, green bacteria), which perform photosynthesis without producing oxygen. It is curious how a green bacterial enzyme was introduced into plants and algae. No homolog of DGD1 has

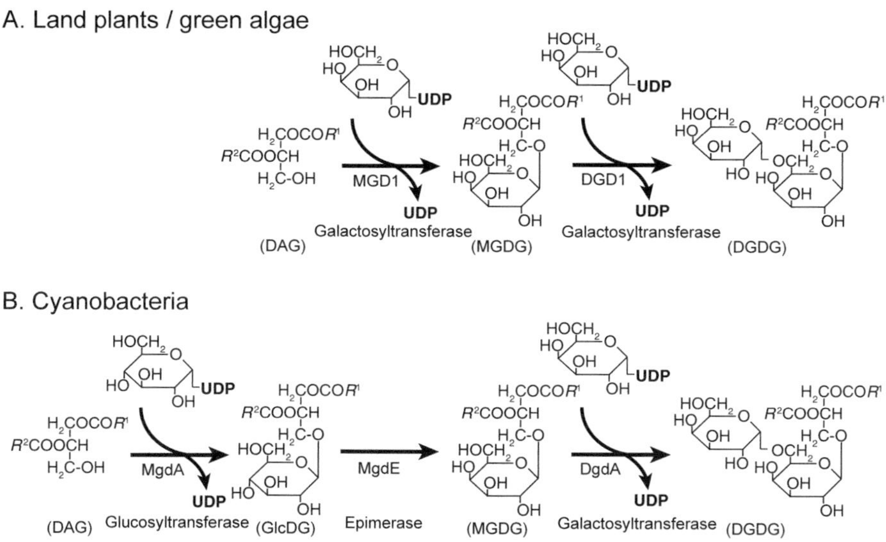

Fig. 7.29 Glycolipid synthesis in chloroplasts and cyanobacteria. See text for details. Original drawing from the Japanese version

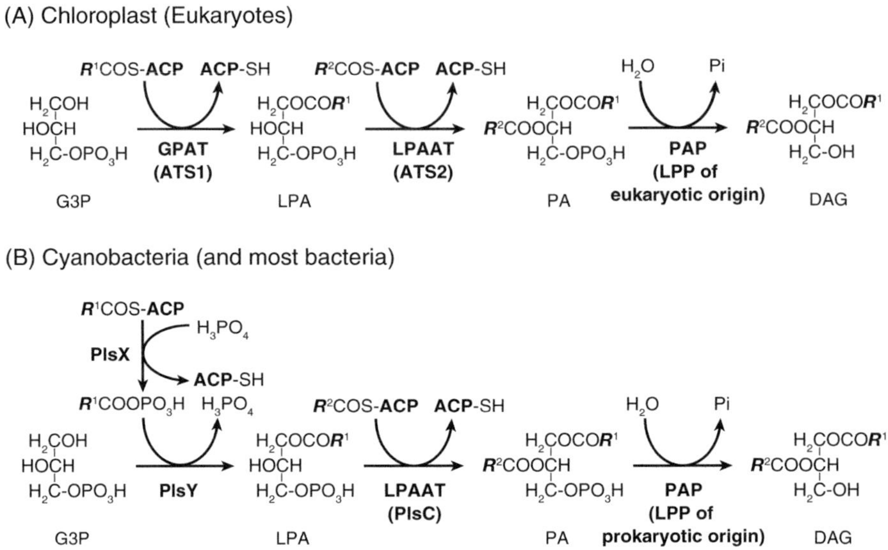

Fig. 7.30 Synthesis of diacylglycerol in chloroplasts and cyanobacteria. See text for details. Reproduced from Sato and Awai 2017 with permission of the publisher

been found in non-photosynthetic organisms, and the origin is still unclear (note that some bacterial homologs of DGD1 are found in the database, but their sporadic occurrence suggests that they are not origins but results of horizontal gene transfer from the plant/algal *DGD1* gene). Some red algae such as *Cyanidioschyzon merolae* in the Cyanidiales lack DGD1, but retain cyanobacteria-type DgdA encoded in the plastid genome. On the other hand, DgdA is not found in marine red algae, which contain DGD1 encoded by the nuclear genome. Note that DGD1 and DgdA catalyze the same biochemical reaction (galactosylation of MGDG), but they belong to different families of glycosyltransferases. We have to assume that ancestral red algae had both DgdA and DGD1, in other words, that DGD1 was acquired in the common ancestor of red algae and green algae.

It is curious that both cyanobacteria and plant chloroplasts have different pathways for synthesizing similar glycolipids, using enzymes of different origins. In addition, MGD1 and DGD1 are both encoded by the nuclear genome in plants and algae.

7.2 Other Enzymes for Lipid Biosynthesis

In collaboration with Koichiro Awai, Shizuoka University, I analyzed enzymes involved in the synthesis of DAG, as well as other lipid synthesis enzymes (Sato and Awai 2017) (Fig. 7.30). The first acyltransferase that transfers an acyl group (namely, fatty acid residue) to glycerol 3-phosphate (G3P) is fundamentally different in prokaryotes and eukaryotes (including plastids). In prokaryotes, the acyl group of acyl ACP (ACP stands for acyl-carrier-protein), the final product of fatty acid synthesis, is transferred to phosphate to form acyl phosphate, which acts as a direct acyl donor to G3P. In plastids, ATS1 transfers acyl group from acyl ACP to G3P. Another acyltransferase exists in endoplasmic reticulum (ER) of eukaryotic cells.

Lysophosphatidic acid (LPA), product of the first acylation, is acylated by the second acyltransferase, a member of a large protein family called LPAAT. Plastid ATS2 and cyanobacterial PlsC are both members of LPAAT, but phylogenetically distant. Phylogenetic analysis of LPAAT has been difficult, because tree form changes depending on added taxa (Sato and Awai 2017). But the most probable results suggested that the clade of eukaryotic LPAATs including plant ATS2 and yeast SLC1 is isolated from the clade of prokaryotic (including cyanobacterial) PlsC. The LPAATs in the plant ER called LPAT belong to a further different clade.

The phosphatase that removes a phosphate group from phosphatidic acid (PA) to form DAG is a member of the family LPP (lipid phosphate phosphatase), which consists of two large clades, each consisting of prokaryotic and eukaryotic enzymes, respectively. Plastid LPP diverged from eukaryotic enzymes. In summary, the plastid pathway for the synthesis of DAG from G3P is not related to the corresponding pathway in cyanobacteria: in other words, the plastid pathway of DAG synthesis did not originate from cyanobacteria (Sato and Awai 2017).

After the publication of the abovementioned papers, I continued to analyze phylogenetic relationship of various other enzymes involved in chloroplast lipid biosynthesis. Phosphatidylglycerol (PG) is synthesized from PA via CDP-DG and PGP (Fig. 7.31). The enzyme CDS that transfers a CMP group from CTP to PA to form CDP-DG is present in both ER and plastids. Plastid isozymes CDS4 and CDS5 are a sister group of cyanobacterial CdsA enzymes, suggesting that they originate from cyanobacteria or an ancestor of cyanobacteria. Plastid PGPS that synthesizes phosphatidylglycerol phosphate (PGP) originated from γ-proteobacteria, whereas plastid PGPP that de-phosphorylates PGP to form PG is a eukaryotic enzyme.

SQDG is synthesized by transfer of sulfoquinovose from UDPsulfoquinovose to DAG. Two different types of enzymes called SqdX and SqdD, respectively, catalyze this reaction. Some cyanobacteria and purple photosynthetic bacteria (α-proteobacteria) possess SqdD, whereas plants, algae, and many cyanobacteria use SqdX/SQD2 (different names in prokaryotes/plants-algae). The latter is a distant homolog of DgdA. It is likely that SqdX was acquired by cyanobacteria from green bacteria. Plant SQD2 originated from cyanobacteria or green bacteria, or their common ancestor (Sato and Awai 2016; Sato 2019).

After examining all enzymes involved in the synthesis of plastid membrane lipids, we recognize that most of them are not heritage from cyanobacteria. The only enzyme that clearly originates from cyanobacteria is DgdA in Cyanidiales red algae. CDS4/5 and SQD2 are likely to originate from an ancestor of cyanobacteria, namely, another, extinct lineage that diverged from extant cyanobacteria. These indicate that popular schemes explaining cyanobacterial origin of chloroplast (Introduction, Fig. 1.9) are no longer valid in their present form. If a cyanobacterial endosymbiont enters a host cell, the enzymes that synthesize the visible structural elements or

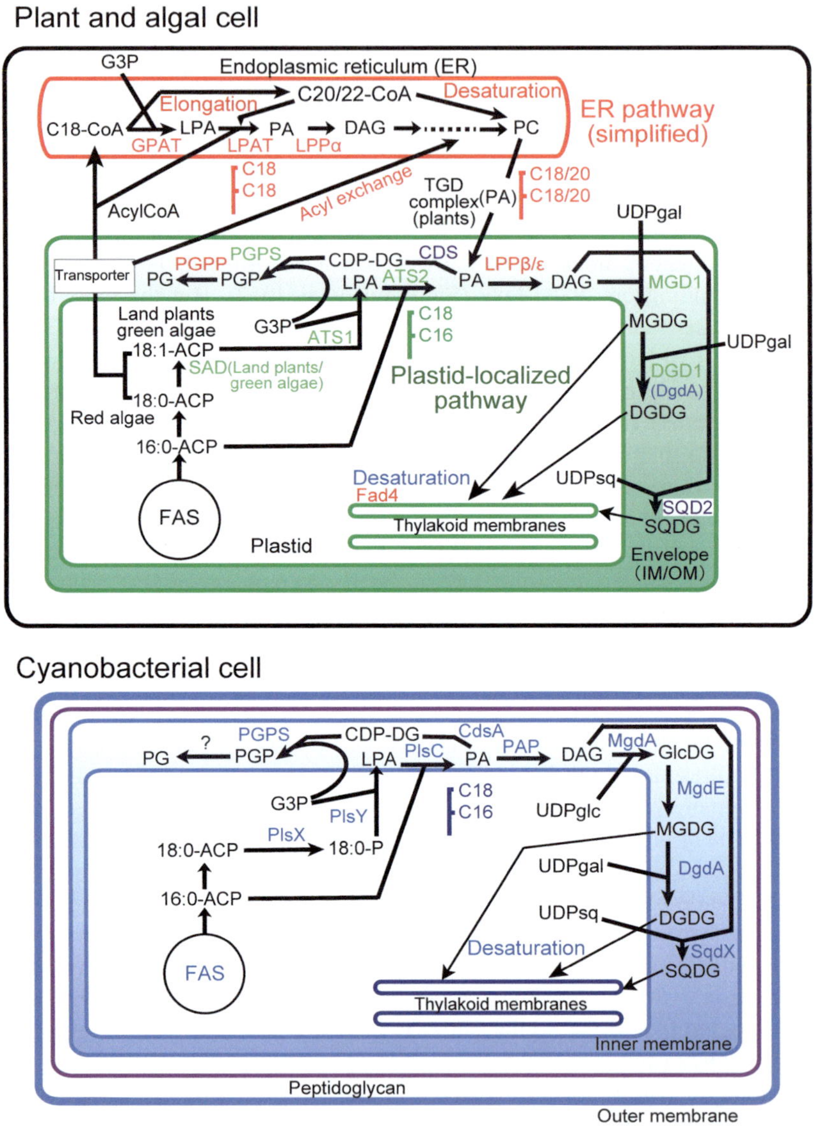

Fig. 7.31 Pathways of lipid biosynthesis in plants/algae and cyanobacteria. (Top) Lipid biosynthesis in plants and algae emphasizing the localization within the cell. Gene-based enzyme names are color-coded according to the phylogenetic types defined in the text and Fig. 7.33. Blue, Type 1; purple, Type 2; red, Type 3; green, Type 4. See text for details. ACP, acyl carrier protein; G3P, glycerol 3-phosphate; LPA, lysophosphatidic acid; PA, phosphatidic acid; DAG, diacylglycerol; MGDG, monogalactosyl diacylglycerol; DGDG, digalactosyl diacylglycerol; UDPgal, UDP galactose; UDPsq, UDP sulfoquinovose; SQDG, sulfoquinovosyl diacylglycerol; CDP-DG, CDP diacylglycerol; PGP, phosphatidylglycerol phosphate; PG, phosphatidylglycerol; FAS, fatty acid synthetase; CoA, Coenzyme A; PC, phosphatidylcholine; IM, inner membrane; OM, outer membrane. (bottom) Lipid biosynthesis in cyanobacteria. All enzyme names are colored blue. UDPglc, UDP glucose; GlcDG, monoglucosyl diacylglycerol. Reproduced from Sato (2019) with permission of the publisher

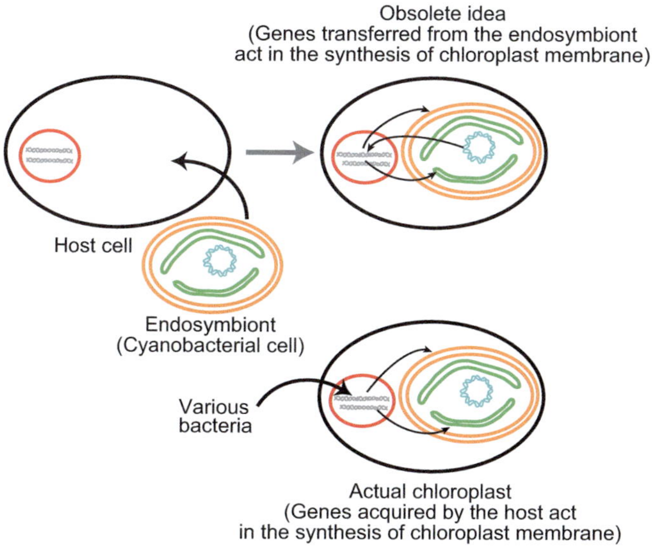

Fig. 7.32 Two ideas on the origin of membrane biosynthesis genes. Obsolete idea presumed that the cyanobacterial endosymbiont genes transferred to the nucleus are involved in the synthesis of chloroplast membrane lipids. Recent phylogenetic analysis as described in the text revealed that the genes acting for the synthesis of chloroplast membrane lipids did not originate from cyanobacteria. Original figure from the Japanese version

membranes are actually provided by the host cell rather than the endosymbiont (Fig. 7.32). I am afraid that researchers still believe that the genes encoding all lipid-synthesizing enzymes that were introduced along with the endosymbiont at the initial endosymbiotic event were transferred to the nucleus and continued to contribute to the synthesis of endosymbiont membranes (endosymbiotic gene transfer or EGT). This is an obsolete idea. One idea to explain the actual situation of chloroplast could be that these nuclear-encoded enzymes were replaced by exogenous enzymes one by one after the formation of chloroplast. In other words, a chromatophore-like endosymbiont (a chromatophore is an endosymbiotic organelle of *Paulinella*) was gradually transformed into chloroplasts/plastids. Transition of chromatophore to chloroplast is an attractive hypothesis, but the actual situations of these two are quite different. Chromatophore genome encodes all enzymes involved in the synthesis of chromatophore membrane lipids (Sato and Awai 2016, 2017; Sato 2019). I suppose that the exogenous genes for lipid-synthesizing enzymes could have been acquired by the host cell before acquisition of cyanobacterial ribosomes (or primary endosymbiosis). I will explain this new idea in the next chapter after showing phylogenetic trees of various other plastid enzymes.

To facilitate understanding of readers, I provide a figure summarizing all my phylogenetic analyses of plastid enzymes (Fig. 7.33), in which four types of trees are presented. Note that the term “plastid enzyme” is used here to mean an enzyme localized and acting in the plastid, rather than plastid-encoded enzyme. “Plastid” is a more general term than “chloroplast.”

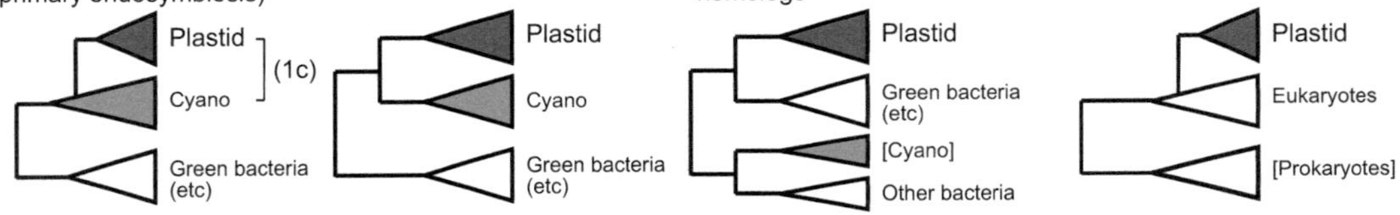

Type of phylogenetic tree	(1) Plastid proteins originated from cyanobacteria (potentially by primary endosymbiosis)	(2) Plastid proteins originate from an ancestor of cyanobacteria	(3) Plastid proteins are not related to cyanobacterial proteins, or no cyanobacterial homologs	(4) Plastid proteins originate from eukaryotes
Membrane lipid synthesis	DgdA*, CDS4/5	SQD2, SQD1[G]	ATS1, ATS2, MGD1, DGD1, PGPS[RG], SQD1[R]	PGPP, LPP
Fatty acid synthesis	AccA[RG], AccB[RG], AccD*, ACP*[R], ACP[G], FabG, FabI[R], FabH*[R], FAD6, FAD7	AccC[RG], FabZ, FabI[G], FabH	FabD[RG], FabF, SAD	FAD4
Peptidoglycan synthesis	MurA, PBP*[some greens]	MraY	MurB, MurC, MurD, MurE, MurF, MurG, Ddl, PBP	
Division	MinD	FtsZ	MinE	
Gene expression	Ribosomal proteins*, rRNA*, RpoA*, ((GyraseA))	GyraseB, DnaB*[R], DnaG*[R], TOP1[R], ((RpoB*, C1+C2*))	RNaseH, TOP1[G], Pol I 5'→3' exo	DNA polymerase (POP), SSB
Photosynthesis	NdhA*, B*, F*, H*, K* ((NdhG*, I*)), HemY (1c) Photosynthesis proteins (*)	NdhC*, E* [Streptophytes]		
Carbon fixation	RbcL*[G], TKL, GAPDH, PRK	PGK, RPE	RbcL*[R]	TPI, FBA, FBP, SBP, RPI

(5) Cyanobacterial proteins having no homologs in plants and algae: MgdA (except the one in *C. merolae* having no activity), MgdE

Fig. 7.33 Classification of phylogenetic relationships between plastid enzymes and cyanobacterial homologs. [G] and [R] indicate, respectively, green lineage (land plants + green algae) and red lineage (red algae, Cryptophytes, Stramenopiles). [RG] indicates that the two lineages are separated, but both present an identical tree type. Asterisk (∗) indicates that the protein is encoded by the Cp genome. Proteins in double parenthesis need special remarks on the tree topology, as follows: NdhG and I, overall tree topology of cyanobacteria is unusual. RpoB and C, both Type 1 and Type 2 topologies are found depending on method. Phylogenetic tree of Gyrase A is complex. There are duplicate clades of cyanobacteria, in which [R] and [G] are inserted separately. This could be Type 2. Cyano, cyanobacteria. Names of enzymes are explained in the text. Revised from Sato (2019) with extensive additions

Type 1: Plastid enzymes diverged from cyanobacterial homologs. This represents the phylogenetic relationship of plastid enzymes that originated from cyanobacteria, such as rRNA and many enzymes encoded by the plastid genome (Chap. 6, Figs. 6.21, 6.22, and 6.23). Type 1c is a related tree for photosynthesis-related enzymes conserved in only cyanobacteria and chloroplasts.

Type 2: Plastid enzymes are a sister group of cyanobacterial homologs. This type has been confused with Type 1 in previous studies, and used to support cyanobacterial origin of plastid enzymes. Because this type of phylogenetic trees was estimated for many enzymes, we have to examine this type as a different type from Type 1.

Type 3: Plastid enzymes are not directly related to cyanobacterial homologs. Rather, they originate from other bacteria.

Type 4: Plastid enzymes originate from eukaryotes. Eukaryotic enzymes belong to a clade different from a clade of prokaryotic enzymes.

Another type should also be added to complete comparison of plastids and cyanobacteria.

Type 5: Cyanobacterial enzymes have no homologs in plants and algae.

Although we will examine the phylogenetic patterns in detail in the following sections, Fig. 7.33 shows that Type 2 trees are abundant in enzymes involved in fatty acid synthesis and DNA replication. The origin of these plastid enzymes is not found in the current cyanobacterial lineage, but in an ancestor of cyanobacteria. It is true that we have to be cautious with potential errors in phylogenetic estimation, but a number of Type 2 trees with different methods for different enzymes might suggest that at least some enzymes originated from an ancestor of cyanobacteria. Is this possible without time machine? We will come back to this problem in the next chapter.

7.3 Enzymes Involved in Fatty Acid Synthesis

Fatty acid synthesis provides acyl groups necessary for lipid biosynthesis. It is a condensing process of acetyl CoA (CoA stands for coenzyme A), an activated form of acetic acid. Acetyl CoA is explained in biochemistry textbooks as a substance that glycolysis provides for citric acid cycle. In photosynthetic organisms, it is not provided by degradation of glucose, but rather synthesized from triose phosphate, a product of photosynthetic carbon fixation cycle called Calvin-Benson cycle, in a similar way to the downstream half of the glycolytic pathway. For use in fatty acid synthesis, acetyl CoA is further activated by acetyl CoA carboxylase, which adds a carbon dioxide molecule to produce malonyl CoA, which is then converted to malonyl ACP (ACP stands for acyl carrier protein), the actual substrate for fatty acid synthase. The added carbon dioxide moiety of malonyl ACP is removed upon subsequent condensing reaction, which allows incorporation of the original acetyl moiety into the fatty acid product.

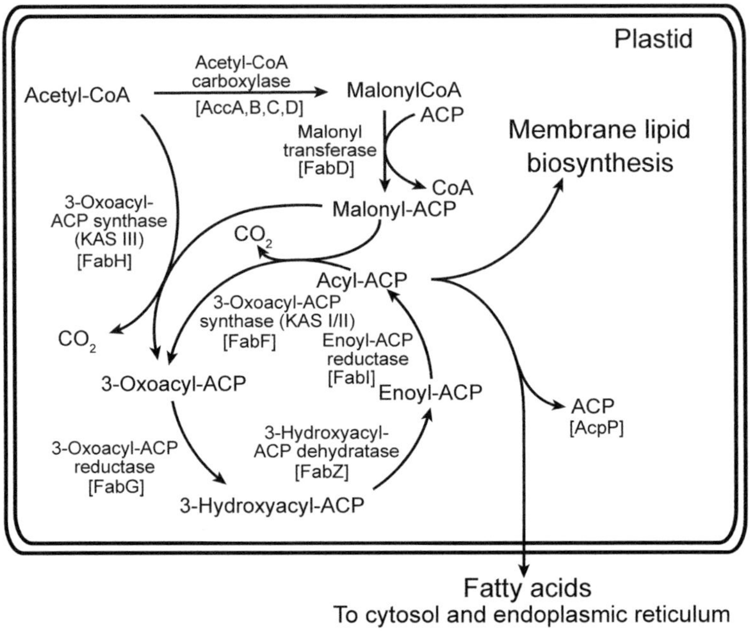

Fig. 7.34 Fatty acid biosynthesis in plastids of most plants and algae. See text for details. Adapted from Sato (2019). Reproduced with permission of the publisher

The initiation reaction of fatty acid synthesis (Fig. 7.34) condenses acetyl CoA and malonyl ACP to form butyryl ACP (C4 acyl group). This reaction is catalyzed by a special condensing enzyme called KAS III, while subsequent condensation reactions are catalyzed by KAS I. The final step elongating from palmitoyl ACP (C16 acyl group) to stearoyl ACP (C18 acyl group) is catalyzed by KAS II. Each of the products of these condensation reactions (3-oxoacyl ACP) bears a carbonyl group at the third carbon from the terminal carboxyl group, which is reduced to a hydroxyl group to form 3-hydroxyacyl ACP. The hydroxyl group is then removed by a dehydratase to form enoyl ACP, containing a double bond. Addition of two hydrogen atoms to the double bond (this is a reduction reaction) yields a saturated acyl ACP, having two additional carbon atoms compared with the original substrate.

Plastid acetyl CoA carboxylase is a prokaryotic enzyme consisting of four subunits called AccA, AccB, AccC, and AccD, respectively. AccD is encoded by the plastid genome in many plants and algae. Plastid genomes of some algae also encode AccA and AccB. In the cytosol of eukaryotic cells (plants, algae, fungi, and animals) resides another type of acetyl CoA carboxylase, a large, multifunctional enzyme integrating the functionality of the four subunits of prokaryotic enzymes. Phylogenetic analysis of the four subunits of plastid and cyanobacterial enzymes revealed Type 1 or Type 2 trees (Fig. 7.33). Plastid malonyltransferases (FabD) showing a Type 3 tree did not originate from cyanobacteria. KAS III (FabH) that initiates fatty acid synthesis cycle shows a Type 2 tree.

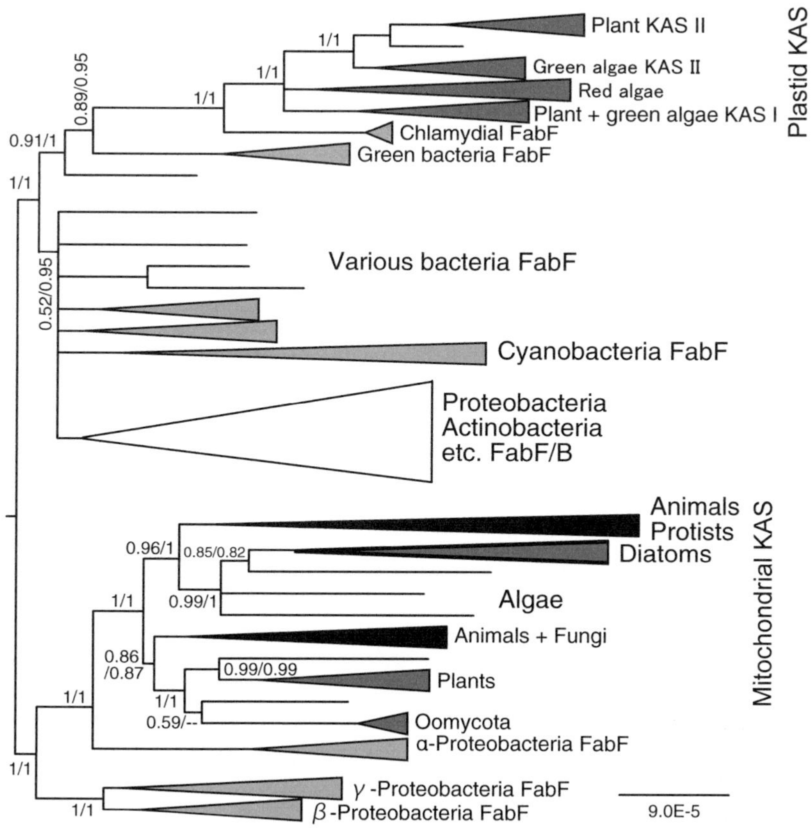

Fig. 7.35 Phylogenetic tree of the condensing enzyme KAS I/II (FabF) acting in fatty acid synthesis. Plastids and mitochondria have distinct condensing enzymes. In mitochondria of both plants and animals, this enzyme is involved in the synthesis of lipoic acid (or short-chain fatty acids). Plastid KAS is related to FabF of Chlamydiae and green bacteria. I suspect that FabF was transferred from green bacteria to both Chlamydiae and plastids. Mitochondrial KAS is monophyletic, and closely related to α-proteobacteria. This is consistent with the idea that mitochondria originated from α-proteobacteria or their related bacteria. Redrawn from the Japanese version

The main cycle of fatty acid synthesis consists of four steps, and hence involves four enzymes. Plastid 3-oxoacyl ACP reductase (FabG) originates from cyanobacteria (Type 1). Plastid 3-hydroxyacyl ACP dehydratase (FabZ) originates from an ancestor of cyanobacteria (Type 2). Plastid condensing enzyme(s) KAS I/II (FabF) is likely to originate from green bacteria, while chlamydial homologs are the most related to the plastid clade (Fig. 7.35). Some species of *Chalamydia* are serious human pathogens. Comparative genomic studies showed that Chlamydiae share some genes with plants, which is taken as evidence for a role of *Chlamydia* during the initial stage of cyanobacterial endosymbiosis (Ball et al. 2015). These authors hypothesized that a cyanobacterial cell and a chlamydial cell coexisted in the ancestor of plant/algal cell as endosymbionts. We will recall Chlamydiae later in the analysis of peptidoglycan. Note that KAS I and KAS II are not differentiated in red algae. Nevertheless, all KAS I/II enzymes are monophyletic (forming a single clade). Acyl carrier protein (ACP) originates from cyanobacteria (Type 1). Complexity is found in the case of enoyl ACP reductase (FabI): Red algal enzymes originate from cyanobacteria, whereas the enzymes in the green lineage originate from an ancestor of cyanobacteria or Chlamydiae (Type 2 or 3).

Desaturases are also important enzymes for fatty acid synthesis. A desaturase introduces a double bond within an acyl group at a precise, fixed position. A delta 9 desaturase creates a *cis* double bond between the ninth and tenth carbons from the carboxyl group. In the plastids of the green lineage, this reaction is catalyzed by stearoyl ACP desaturase (SAD), a soluble enzyme in the stroma. SAD is not present in cyanobacteria, but found in various other bacteria (Type 3 tree). Delta 12 (DesA/FAD6) and delta 15 (DesB/FAD7) desaturases of cyanobacteria/plastids are embedded in the membranes and catalyze introduction of a *cis* double bond in the acyl group within a lipid molecule. Cyanobacterial origin of these desaturases was reported before (Sato and Moriyama 2007). Another plastid desaturase FAD4 introduces a *trans* double bond between the third and the fourth carbons of palmitoyl group at the *sn*-2 position (central carbon of glycerol) of PG. This unique desaturase is a homolog of ubiquitin ligase component E2, and is only found in plants and algae (excluding *Paulinella*), suggesting a Type 4 phylogenetic relationship.

These results show that many of the enzymes involved in fatty acid synthesis originate from cyanobacteria or their ancestor (Type 1 or 2). This may be explained by coordination of multiple subunits in acetyl CoA carboxylase and fatty acid synthase. Functional restriction could be imposed on the evolution of each subunit of a multisubunit complex. We still need studies on this point.

7.4 Plastid Enzymes Involved in DNA Replication, RNA Transcription, and Division

Discontinuity of cyanobacteria and plastids is also evident in the replication enzyme of plastid DNA. According to sequence homology, the plastid DNA itself is undoubtedly acquired from cyanobacterial DNA. Nevertheless, the genome size is at least ten times different: the size of the smallest cyanobacterial genome is about 2 Mbp (mega base pairs), whereas the largest plastid DNA amounts to about 0.2 Mbp (=200 kbp, kilo base pairs).

We focus now on the mechanism of DNA replication. Cyanobacteria, like other bacteria, have DNA polymerase III, which is a highly processive (or efficient in plain word) enzyme with many subunits, and suited for replication of an entire bacterial genome. In contrast, plastid DNA is replicated by a DNA polymerase encoded by the nuclear genome, which consists of a single, large polypeptide. This enzyme called POP (plastid and protist organellar DNA polymerase) functions in both plastids and mitochondria. In plants and algae, the POP enzyme encoded by an identical

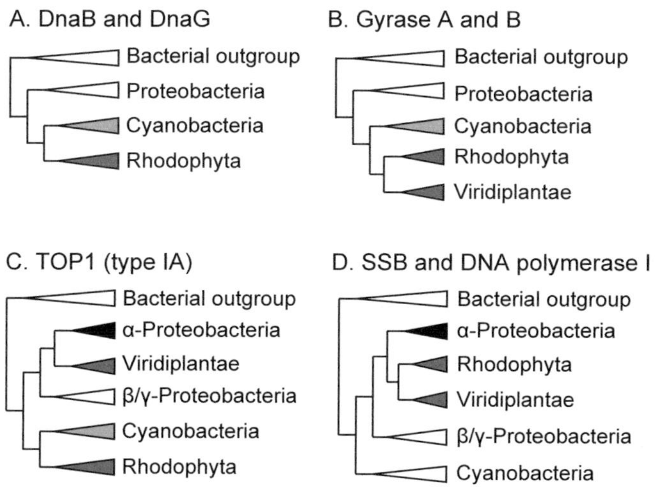

Fig. 7.36 Phylogenetic relationships of enzymes related to organellar DNA replication. Phylogenetic tree of Gyrase A is, in fact, complex. There are duplicate clades of cyanobacteria. Another clade of cyanobacteria is located between Rhodophyta and Viridiplantae, as a sister group to Rhodophyta. This could be Type 2. Reproduced from Moriyama et al. (2014) under Creative Commons CC BY license

gene is synthesized in the cytosol, and targeted to both plastids and mitochondria. In *Arabidopsis thaliana*, two copies of POP genes exist, but the products of both genes are targeted to the two organelles. Phylogenetic analysis indicated that POP was initially a mitochondrial replication enzyme, and then retargeted to plastids after the formation of plastids. In vertebrates, insects, and fungi, unrelated DNA replication enzyme is present in mitochondria. It was discovered as the third DNA polymerase and called gamma (γ) in early days of molecular biology. Many biochemistry textbooks still describe that gamma is a mitochondrial replicase in all eukaryotes, but the fact is that vast majority of eukaryotic organisms including protists such as *Tetrahymena* use POP rather than gamma. Probably, ancestral eukaryotes acquired POP of a viral origin soon after the formation of mitochondria, for use in the replication of mitochondrial DNA. In an ancestor of Opisthokonta (animals and fungi), mitochondrial replication enzyme was switched from POP to gamma (Moriyama and Sato 2014). Note that most protists belong to Bikonta and retain POP. This tale of DNA replicase may be understood as an interesting example of "creating animals from plants" as imagined by supporters of the autogenous theory. Phylogenetic origin of POP is, therefore, Type 4.

Figure 7.36 shows schematically the phylogenetic relationships of enzymes related to plastid DNA replication, such as DNA helicase (DnaB, DnaG), DNA gyrase, DNA topoisomerase, and single-stranded DNA-binding protein. Red algae retain some cyanobacterial components, but a simple rule is not evident.

Plastid genome encodes a prokaryotic RNA polymerase consisting of five types of subunits, α, β, β', β'', and σ. In a single polymerase complex, there are two molecules

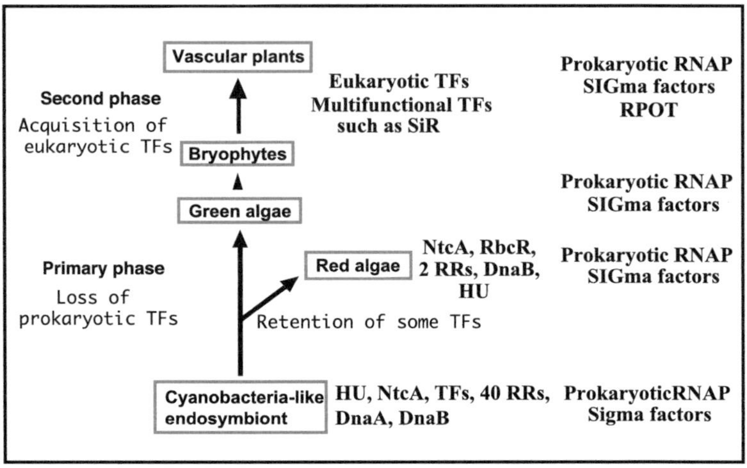

Fig. 7.37 Discontinuous evolution of plastid genomic machinery. See text for explanation. RNAP, RNA polymerase; RPOT, phage-type RNA polymerase; RR, response regulator; TF, transcription factor. NtcA, RbcR, DnaB, and NtcA are examples of prokaryotic transcription factors encoded by the plastid genome in the red lineage. SiR, SIGma factors, and RPOT are encoded by the nucleus. Modified from the Japanese version based on Sato (2001, 2002J)

of α subunit, whereas other subunits are present as a single copy. The β' and β" subunits are combined as a single polypeptide called β' in many bacteria, but they are distinct polypeptides (encoded by separate genes) in cyanobacteria and plastids. In a red alga *Cyanidioschyzon merolae*, however, the two subunits are fused. In all plants and algae, the σ subunit is a small protein encoded by the nuclear genome, and several different *SIG* genes are differentially expressed depending on different environmental conditions. The plastid genes for the α, β, β', and β" subunits are called *rpoA*, *rpoB*, *rpoC1,* and *rpoC2*, respectively. Phylogenetic analysis of the RpoA, B, C1, and C2 proteins (these are alternative names of the subunits based on gene names) were performed, and a Type 1 tree was obtained for RpoA. Plastid RpoC1 and RpoC2 were combined to construct a tree with bacterial RpoC proteins. Various different shapes of trees were obtained with RpoB and RpoC1 + C2, but I tentatively identified Type 2 trees for them (Sato 2019).

Gene expression system is radically different in plastids and cyanobacteria. The theory of discontinuous evolution of plastid genomic machinery that I proposed before (Sato 2001) tentatively accepted cyanobacterial origin of plastids, and estimated the fate of various transcription factors during the evolution (Fig. 7.37): it should be noted that only a few cyanobacterial enzymes or proteins were transmitted to ancestral plastids, and retained by red algal plastids. Essentially no cyanobacterial proteins (except RNA polymerase) were transmitted to the green algal ancestor, while many new transcription factors and nuclear-encoded RNA polymerase (NEP) were provided from the eukaryotic host. Overall, this scenario emphasizing the discontinuity of plastids from cyanobacteria is roughly consistent with the scenario of lipid biosynthesis pathways. However, the initial loss of most components and later acquisition of replacement components in transcription factor scenario is different from the scenario of lipid-related enzymes that were replaced early in the ancestral plastids. The scheme in Fig. 7.37 is already an old version of my theory, and will have to be revised in a near future. For example, if all plastid transcription factors were lost in the ancestral algae, gene regulation would not be possible. Replacement factors could have been prepared before the formation of plastids.

Prokaryotic division proteins MinD and MinE are known to act in the plastid. A Type 1 tree was obtained for MinD, whereas a Type 3 tree was found for MinE (Sato 2019). FtsZ, another protein involved in chloroplast division, showed a Type 2 tree in both my analysis Sato (2019) and Grosche and Rensing (2017).

7.5 Origin of Plastid Peptidoglycan

Bacteria are classified into Gram-positive and Gram-negative, depending on reactivity to Gram staining reagent. Gram-positive bacteria include Firmicutes (*Bacillus subtilis* is a well-known model bacterium of this group) in which each cell is covered by a thick layer of peptidoglycan. Each cell of Gram-negative bacteria is surrounded by outer and inner membranes and a peptidoglycan layer located between the two membranes. Peptidoglycan is a high molecular weight, mesh-like polymer consisting of a rare amino acid, D-alanine, and various sugars such as muramic acid. Peptidoglycan has not been detected in mitochondria. It is not present in angiosperm chloroplasts, but, as mentioned in Sect. 6.4, it has been known as a covering layer of the cyanelle of *Cyanophora paradoxa* and *Paulinella chromatophora*. That is why these cyanelles have long been believed to represent typical examples of cyanobacterial cells actually living inside a eukaryotic host, in other words, a principal experimental evidence that supported the endosymbiotic origin of chloroplast.

Many enzymes are involved in the synthesis of peptidoglycan. One category of enzymes among them is known as penicillin-binding proteins (PBP). Penicillin and other β-lactam antibiotics inhibit the activity of PBP to prevent formation of the mesh-like structure of peptidoglycan, and thus facilitate osmotic lysis of actively growing bacteria.

When deciphering of genomes of model organisms began around the year 2000, the genome of a moss *Physcomitrella patens* was sequenced. Professor Hiroyoshi Takano and his colleagues in Kumamoto University discovered various peptidoglycan synthesis enzymes in this moss, and they identified all the genes that encode the enzymes of peptidoglycan synthesis pathway. They found that disruption of each of these genes prevented chloroplast division, thus resulting in cells containing a single, large chloroplast (Takano and Takechi 2010). This was curious because a normal cell contains about 50 chloroplasts. Nevertheless, nobody has ever observed a peptidoglycan layer in the chloroplasts of any mosses. Researchers considered that peptidoglycan is necessary for plastid division, but that the peptidoglycan layer might be very thin, or temporarily accumulates during the division in moss chloroplasts. A complete set of peptidoglycan synthesis enzymes was also identified in a green alga

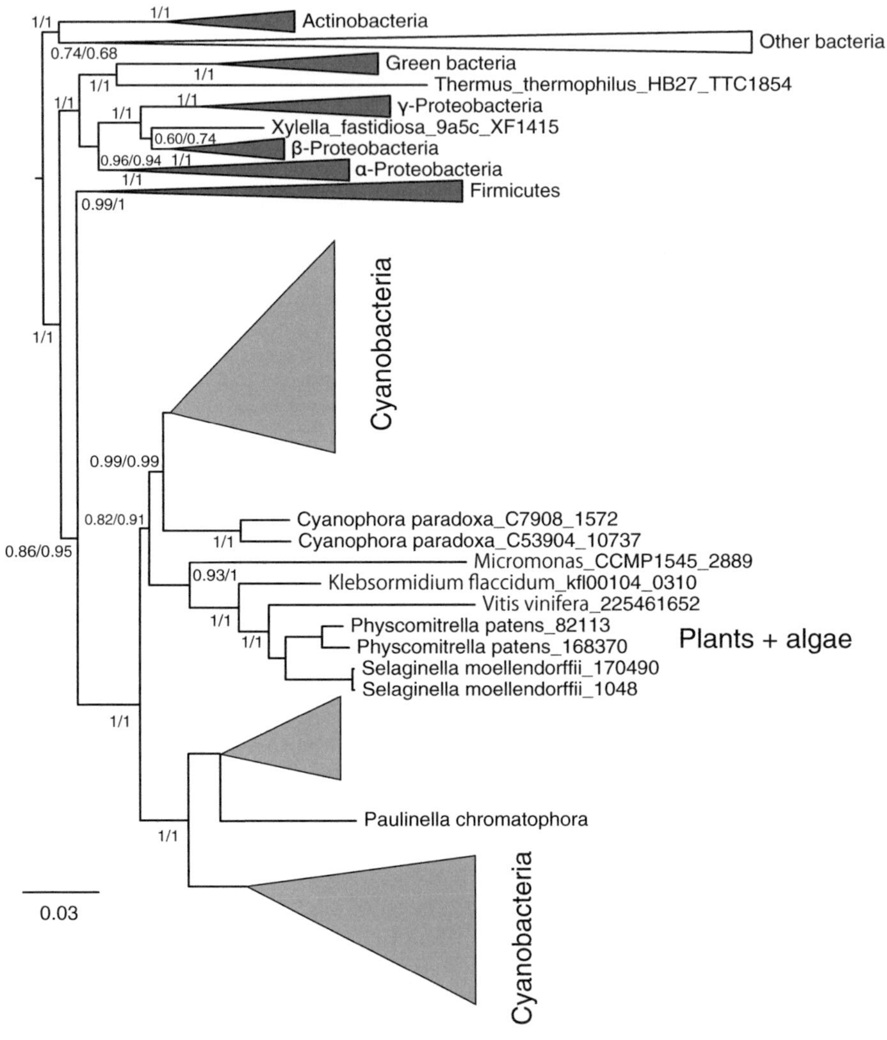

Fig. 7.38 Phylogenetic tree of MurA. The cyanobacterial clade is enlarged. This enzyme is present in various bacteria. The clade of plants and algae diverged from inside of the cyanobacterial clade. Note that *Cyanophora* enzyme and plant/algal enzymes are not monophyletic. Modified from Sato and Takano (2017). Reproduced with permission of the publisher

Micromonas spp., a fern *Selaginella moellendorffii*, a land green alga *Klebsormidium flaccidum*, and some gymnosperms (Table 7.10). Angiosperms such as *Arabidopsis thaliana* also contain the genes for some of these enzymes, but disruption of them did not result in notable defects. The angiosperm homologues might be involved in other chloroplast functions other than the synthesis of peptidoglycan.

In 2016, Takano group published interesting results of research using a special staining method with labeled D-alanine that suggest the existence of a peptidoglycan

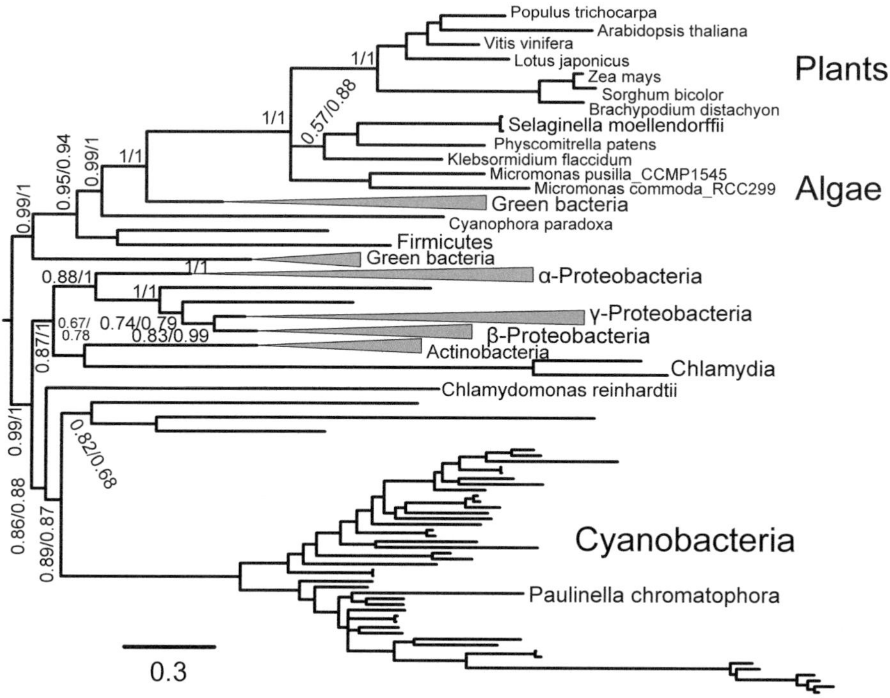

Fig. 7.39 Phylogenetic tree of MurG. MurG protein is present in various bacteria. In this figure, the clades of cyanobacteria and plants/algae are enlarged. These two clades are located at different positions in the tree. Cyanobacteria are not origin of MurG in plants/algae (chloroplasts). Modified from Sato and Takano (2017). Reproduced with permission of the publisher

layer covering the moss chloroplast. Then, in collaboration with Takano group, we used conventional transmission electron microscope to show that the electron density of the intermembrane space between the two envelope membranes of chloroplast was significantly low in the cells treated with ampicillin, or in the mutant cells defective in peptidoglycan synthesis enzymes compared to wild-type cells (Sato et al. 2017). Therefore, peptidoglycan (or a related substance) is estimated to localize in the intermembrane space.

Then, an important question is asked: is the moss peptidoglycan evidence for the endosymbiotic origin of chloroplasts?

We performed detailed phylogenetic analysis for each of the ten enzymes and found that MurA (Fig. 7.38) and MraY in the green algae and land plants originated from cyanobacteria or an ancestor of cyanobacteria (Type 1 and Type 2, respectively) (Sato and Takano 2017). Cyanobacterial origin was rejected for all other enzymes of green algae and land plants, although their monophyletic relationship was confirmed (Table 7.10). A MurG tree is shown as an example (Fig. 7.39). This suggests that most enzymes of peptidoglycan synthesis in green algae and land plants did not originate from cyanobacteria. Furthermore, different sister groups

Table 7.10 Origin of enzymes involved in peptidoglycan biosynthesis (Sato and Takano 2017)

Symbol	Enzyme name	Occurrence[a,b]	Origin
MurA	UDP-*N*-acetylglucosamine 1-carboxyvinyltransferase	-CKMPRSV	Cyanobacteria
MurB	UDP-*N*-acetylenolpyruvoylglucosamine reductase	-CKMPRSV	Green bacteria
MurC	UDP-*N*-acetylmuramate – L-alanine ligase	--KMPRS-	Actinobacteria
MurD	UDP-*N*-acetylmuramoyl-L-alanyl-D-glutamate synthetase	-CKMPRS-	Chlamydiae
MurE	UDP-*N*-acetylmuramoylalanyl-D-glutamate–2,6-diamino pimelate ligase	ACKMPRSV*	Firmicutes
MurF	UDP-*N*-acetylmuramoylalanyl-D-glutamyl-2, 6-diaminopimelate–D-alanyl-D-alanyl ligase	-CKMP-SV	Chlamydiae and α-Proteobacteria
MraY	Phospho-*N*-acetylmuramoyl-pentapeptide transferase	A-KMPRSV*	Cyanobacteria
MurG	*N*-Acetylglucosaminyl transferase	ACKMPRSV*	Green bacteria
Ddl	D-Alanylalanine synthetase	ACKMPRSV*	α-Proteobacteria
PBP	Penicillin-binding protein (class A)	-CKMPRS-	γ-Proteobacteria

[a]Species name in the list of occurrence: A, *A. thaliana*; C, *C. paradoxa*; K, *Klebsormidium flaccidum*; M, *Micromonas* sp. CCMP 1545; P, *P. patens*; R, *P. chromatophora* (for rhizopod); S, *Selaginella moellendorffii*; V, *Vitis vinifera*

[b]Asterisk indicates that the enzyme was detected in more other species

Original table, enhanced from Sato and Takano (2017)

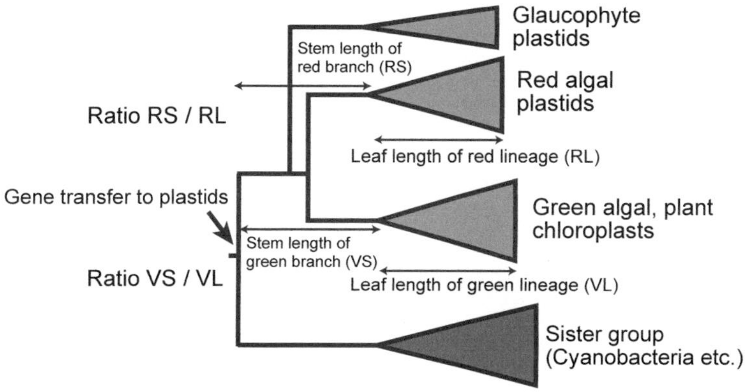

Fig. 7.40 Estimation of stem length. This diagram shows a typical phylogenetic tree of chloroplast proteins. Time goes from left to right. We will use the ratio VS/VL or RS/RL as relative stem lengths for the green and red lineages, respectively. Modified from Sato (2019). Reproduced with permission of the publisher

(most related clade) were found for different enzymes, suggesting diverse origins of the enzymes. The presumptive presence of peptidoglycan in the chloroplasts of some green algae and land plants, therefore, is a result of massive horizontal gene transfers from various bacteria, rather than cyanobacterial origin of chloroplasts. Nevertheless, researchers tend to interpret the results of phylogenetic analysis in a different way, in accordance with the orthodox endosymbiotic theory: Namely, all enzymes in peptidoglycan synthesis were introduced into the proto-algal cell at the time of primary endosymbiosis. Then, all components except MurA and MraY were replaced. We will have to analyze the results in more detail.

7.6 Estimation of the Time of Gene Acquisition

A close examination of the phylogenetic tree of MurA (Fig. 7.38) indicated a very short branch at the root of plants and algae. I tried to estimate the time of acquisition of exogenous genes, according to the method of Pittis and Gabaldón (2016), who showed that the gene transfers from bacteria other than α-proteobacteria preceded the acquisition of mitochondria by comparing the times of gene transfers from various prokaryotes to the eukaryotic ancestor. If an ancestral gene was transferred to a eukaryotic host and evolved as a plant or algal gene, the stem length is a measure of this evolutionary distance (evolutionary time) (VS or RS in Fig. 7.40). This length can be normalized by the evolutionary distance within the green lineage (VL or RL), if evolutionary rate was constant. We will use the normalized stem length, namely, the ratio VS/VL or RS/RL, as a measure of the time elapsed from the gene acquisition to the last common ancestor of green or red lineage.

Table 7.11 Summary of tree topology and estimation of relative stem lengths

Protein or RNA	Tree topology	Green stem (VS/VL)		Red stem (RS/RL)	
rRNA (reference point)	**1**	**0.49**		**0.66**	
MurA	1	0.27	S	–	
MraY	2	0.19	S	–	
MurD	3	1.00	L	–	
MurG	3	0.66	(L)	–	
ATS2	3	1.34	L	0.43	
MGD1	3	1.04	L	1.07	L
SQD2 (SqdX)	2	0.88	L	1.34	L
SQD1 (SqdB)	2/3	0.62	(L)	0.93	L
CDS	1	0.77	L	0.76	
PGPS	3	0.62		0.30	S
KAS I/II	3	1.61/1.01	L	0.54	L
KAS III	2/1	0.26	S	0.35	S
FabD	3	0.87	L	1.18	L
FabG	1	0.71	L	0.39	S
AccA	1	0.47		0.44	S
AccD	1	0.55		0.31	S
FtsZ1/3 (green) or FtsZ (red algae)	2	0.35	S	0.31	S
RbcL	1 +	1.04	L	0.76	
PsaA/PsaB	1c	0.75/0.72	L	0.61/0.72	
23 proteins of photosynthesis	1c	0.85	L	1.01	L
ChlB/N/L	1 +	0.66	(L)	–	
33 conserved proteins	1	0.54		0.65	
Ndh proteins	1	0.42		–	

–, no homolog present. In phylogenetic pattern,“1 +” indicates that cyanobacterial proteins are polyphyletic. L, markedly long stem; S, markedly short stem. Revised from the original Japanese version

Table 7.11 lists some results of estimation of relative stem lengths. The values obtained by the ML and BI methods were similar. If, for convenience of explanation, acquisition of ribosomal RNA gene is defined as a reference time point (many researchers take this point as the primary endosymbiosis by a cyanobacterial cell), we obtain $[VS/VL]_{ref} = 0.49$ and $[RS/RL]_{ref} = 0.66$. These can be regarded as values for the genes introduced at the time of the primary endosymbiosis in the green and red lineages, respectively. The two values may be expected to coincide, but the actual difference could represent the differences in positions of the last common ancestors in the two lineages. Conserved proteins including ribosomal proteins gave similar values in respective lineages, suggesting that ribosome as a whole was a product of the primary endosymbiosis. Very small values (marked “S”) were obtained for MurA and MraY, suggesting that these were acquired from cyanobacteria, but after the acquisition of rRNA (or the primary endosymbiosis). The origin of *Cyanophora* MurA is likely to be slightly different (Fig. 7.38), suggesting that MurA was acquired independently in glaucophytes and green algae. A late acquisition is also estimated for MraY, in which no glaucophyte homolog was obtained (Table 7.10). It is interesting to note that cyanobacterial gene transfer occurred several times, one with rRNA and the other with MurA and MraY. The VS/VL values for MurD, MurG, and Ddl (not shown) were quite large, suggesting that the acquisition of these components preceded the primary endosymbiosis (acquisition of rRNA). In contrast, the values for MurB and MurC were small, and the values for MurE, MurF, and PBP were similar to the reference (rRNA). All these results suggest that the acquisition time was variable depending on genes.

Alternatively, evolutionary rate changed drastically upon the primary endosymbiosis. It is unlikely that this can explain the highly divergent values of VS/VL for the members of peptidoglycan system. Selection pressure is expected to be similar for the enzymes of the same metabolic pathway. Here, we have two extreme hypotheses. In the first hypothesis, if we accept that the primary endosymbiont provided all the enzymes necessary for synthesizing all chloroplast structures, including membranes and peptidoglycan, then we have to explain diverse origins of the enzymes involved in the synthesis of these structures by horizontal gene transfer that replaced the original, cyanobacterial enzymes. The large and small values of VS/VL should be explained by variation in evolutionary rates or method of phylogenetic estimation. This is a traditional hypothesis of biologists, expressed implicitly or explicitly. In the second hypothesis, chloroplast was formed as a result of many gene transfers, including the acquisition of rRNA and many other enzymes. The time and source of gene acquisition might be variable, and multiple gene transfers could occur for a single gene or enzyme. In this "relaxed" idea, we can affirm that rRNA and some particular enzymes were acquired by a single gene transfer event, if we have enough evidence. We can choose any position between the two extremities. This will be the major point to be discussed in the next chapter.

We continue to examine the relative stem lengths for other enzymes. Table 7.11 included data in which one of the two lineages presented a large or small value compared to the reference value. A more comprehensive data is available in a recent paper (Sato 2019). Some enzymes of lipid and fatty acid biosynthesis (MGD1, SQD2, SQD1, KAS I/II, and FabD) showed larger values of VS/VL and RS/RL than the references, and were possibly acquired before the acquisition of rRNA. KAS III showed markedly smaller VS/VL and RS/RL values. Similar results were obtained with FtsZ, a protein involved in chloroplast division. KAS III and FtsZ have been considered heritages of cyanobacterial endosymbiont, but we have to change this view: they are likely to be acquired after the acquisition of rRNA, or the primary endosymbiosis. Currently, I have no explanation for the enzymes of lipid biosynthesis showing conflicting VS/VL and RS/RL values.

Curious results were obtained for photosynthesis-related proteins. RbcL, the large subunit of carbon-fixation enzyme, ribulose 1,5-bisphosphate carboxylase/oxygenase, showed a larger VS/VL value. Red algal RbcL is known to originate from α-proteobacteria, and the original cyanobacterial RbcL (according to the traditional view) has been replaced by horizontal gene transfer in red algae (Tabita

1999; Sato 2006). Phylogenetic analysis with recent data suggests a more complicated situation (Chap. 8). This is indeed a good alert that even the genes encoded by the plastid genome could be replaced during the evolution, and the origin and time of acquisition of plastid-encoded enzymes in the green and red lineages may be different. In addition, PsaA and PsaB, the reaction center proteins in the photosystem I, showed larger values of VS/VL. A combined phylogenetic tree using 23 photosynthesis-related proteins encoded by the plastid genome also showed larger VS/VL and RS/RL values. These results suggest that photosynthesis-related proteins might be acquired from cyanobacteria before the acquisition of rRNA. Differences in phylogenetic trees of rRNA, ribosomal proteins, and photosynthesis-related proteins have been noted for a long time, but they were not seriously argued, or simply interpreted in terms of variation in evolutionary rate. The relative stem length can be affected by uneven evolutionary rate, but this does not seem to affect overall trends except for very limited genes, such as RpoB and RpoC (see Sato 2019 for discussion).

Chlorophyll biosynthesis enzymes ChlB/N/L were only conserved in the chloroplasts of green algae, mosses, ferns, and some gymnosperms. The VS/VL value was larger, suggesting again possibly early acquisition of chlorophyll biosynthesis system. We are not sure whether they were acquired with other photosynthesis-related proteins. Ndh genes that encode the components of NADH dehydrogenase complex in chloroplasts in the green lineage are not present in the basal green algae and any red alga. They are likely to be later acquisition, but the VS/VL value did not clearly support this idea. Nevertheless, we have to find an explanation about the origin of these proteins.

7.7 Relationship Between Chloroplasts and Cyanobacteria in the Light of Membrane Structure and DNA Replication: Continuity and Discontinuity

In this chapter, we have been discussing on membranous structure and DNA replication, as well as other components. Traditional arguments on the cyanobacterial origin of chloroplasts equivocally relied on the presence of DNA and the similarity of membrane structure. Although highly homologous sequences of DNA are found in cyanobacterial and chloroplast genomes, marked differences are found in the enzymes in the machineries involved in the functioning of DNA. Phylogenetic analysis of homologous components suggested that many components of the chloroplast genomic machinery did not originate from cyanobacteria. In addition, entirely different replication enzymes are used in cyanobacteria and chloroplasts. The mitochondrial replication enzyme has been retargeted to chloroplasts. Surprisingly, unrelated DNA polymerases are used in the mitochondria of plants and animals. This is not simply explained by α-proteobacterial origin of mitochondria. We have to consider repeated gene transfers from various sources including viruses.

We found intriguing similarity of membrane lipids in cyanobacteria and chloroplasts, but the enzymes involved in the synthesis of these lipids are different in their phylogenetic origins. Curiously, different reactions account for some steps of the biosynthetic pathways. Considering these facts, questions arise as to the visually self-explanatory schemes on the origin of chloroplasts (Sect. 1.5). When a cyanobacterial cell was engulfed by a host cell in the origin, the mechanism must have already existed in the host cell for the synthesis of the endosymbiont membranes. The host had already acquired various enzymes from various bacteria or eukaryotes before the endosymbiosis. Only a limited number of cyanobacterial enzymes function in the chloroplasts. In addition, we also found some curious cyanobacterial enzymes in the chloroplast that were not introduced by the primary endosymbiosis, namely, that were acquired independently from rRNA and ribosomal components, which are tentatively regarded as legitimate products of the primary endosymbiosis. Visually intelligible schemes that explicitly focus on the origin and inheritance of membranes were repeatedly presented by Cavalier-Smith and other researchers, but these are misleading even for researchers. A typical discussion of this category is a quest for the origin of chloroplast outer envelope, namely, whether it originated from the host or from the endosymbiont. Cavalier-Smith (2000) hypothesized that a phagosome membrane enclosing the endosymbiont was present in the initial stage of endosymbiosis, and he discussed which of the three membranes (phagosome membrane, outer and inner membranes of cyanobacterial endosymbiont) disappeared during the establishment of the organelle. Many researchers have been confused by discussions of this kind. I have to emphasize an important principle: *membranes are not inherited by themselves*. We have to consider the involvement of the elements that the host cell had prepared before the endosymbiosis.

Table 7.12 lists apparent similarities of cyanobacteria and chloroplasts that conceal fundamental dissimilarities. As shown in Table 5.7 (Chap. 5), cyanobacteria and chloroplasts (plastids) share a number of similar characters with respect to substance, structure, and function. However, close examination of the underlying mechanisms that support the similarities uncovered differences in enzymes and their phylogenetic origins. We already presented examples, such as glycolipid biosynthesis, DNA replication machinery, and peptidoglycan synthesis, but we can add more items as evidence for the differences in cyanobacteria and chloroplasts. Many of these have been already explained in various textbooks, but they are not recognized as evidence for the discontinuity of cyanobacteria and chloroplasts. I shall briefly explain them one by one, based on Table 7.12.

Tetrapyrrole biosynthetic pathway (or reaction steps) is similar in cyanobacteria and chloroplasts, but, in some reaction steps, two or three different enzymes are known for a single reaction. They are not homologous, and they are structurally different, but they catalyze the same reaction: in short, they are nonhomologous, isofunctional enzymes (Kobayashi et al. 2014). This is an example of convergent evolution, in which a similar trait of different origins is found in evolutionarily unrelated organisms by a functional constraint. Similarity in biological traits normally implies phylogenetic relationship, but convergent evolution is a similarity found in phylogenetically unrelated organisms,

Table 7.12 Discontinuities in apparent similarities of cyanobacteria and chloroplasts

Classification	Item	Cyanobacteria	Chloroplasts
Substance	Tetrapyrrole synthesis	Nonhomologous isofunctional enzymes in at least three reactions	
	Glycolipid synthesis	Entirely different mechanism (see text)	
	DNA replication	Pol III, bacterial complex	Plant protist organellar DNA polymerase (POP), single polypeptide
Structure	Thylakoid membrane formation	"Biogenesis center" is known	Invagination of envelope, prolamellar body, etc.
	Envelope membrane lipids	Glycolipids	Outer envelope resembles endoplasmic reticulum, containing phosphatidylcholine
	Peptidoglycan synthesis	Most enzymes are of different origins (see text)	
Function	Division mechanism	MinCDE, FtsZ, etc.	No MinC. MinDE, FtsZ, and dynamin, a eukaryotic component
	Protein synthesis	Essentially similar, but there are chloroplast-specific components	
	Transcription enzyme(s)	Prokaryotic enzyme (RpoABC1C2D)	Prokaryotic enzyme encoded by the plastid genome, plus eukaryotic enzyme (RpoT) encoded by the nuclear genome. Sigma factors are nuclear-encoded
	Oxygen evolving complex	PsbOUV. PsbP-like component PsbP' is not involved in oxygen evolution	PsbOPQ. Red algae retain PsbOUV

or similarity by unrelated mechanisms. A good example is the protoporphyrinogen IX oxidase (PPO) that converts protoporphyrinogen IX to protoporphyrin IX by removing two hydrogen atoms. This reaction is catalyzed by HemY in most eukaryotes and some bacteria, but HemG acts in the same reaction in *Escherichia coli* and some bacteria. HemJ is known in many cyanobacteria and α-proteobacteria. HemG, HemJ, and HemY are structurally dissimilar enzymes, but can catalyze the same oxidation reaction. Even in cyanobacteria, the three enzymes are used in different strains. Curiously, closely related *Prochlorococcus marinus* strains use different enzymes, maybe as a result of horizontal gene transfer. The three isofunctional enzymes represent an example of convergent evolution, and the use of different enzymes in related species represents horizontal gene transfer.

The mechanism of formation of thylakoid membranes is still a mystery. In chloroplasts, invagination of envelope membranes was proposed before. Restructuring of prolamellar body (present in etioplasts of dark-grown plants) can account for the formation of thylakoid membranes during greening. Involvement of VIPP1 in vesicular transport from envelope membranes to thylakoid membranes was proposed before, but now this protein is considered to have a different function (Zhang and Sakamoto 2015). In cyanobacteria, thylakoid membranes are not synthesized de novo (thylakoid membranes are always present during the growth of cyanobacteria). Recently, a special structure called "biogenesis center" (or "thylakoid center"), which is located near the inner membrane where thylakoid membranes converge. Biogenesis center is supposed to produce thylakoid membranes (Rast et al. 2015). In the green alga, *Chlamydomonas reinhardtii*, a special region of thylakoid membrane (T-zone) was supposed to function in producing components of Photosystem II. The mechanism of thylakoid membranes is probably different in cyanobacteria and chloroplasts. In addition, similarity in lipid composition of outer envelope membrane and endoplasmic reticulum has been known. The outer membrane of cyanobacteria and chloroplasts might not be a homologous structure.

Comparison of similar functions also gives interesting clues to the discontinuity of cyanobacteria and chloroplasts. Chloroplast division is known to involve both prokaryotic and eukaryotic components. Among the prokaryotic components, FtsZ has been considered a typical cyanobacterial heritage. As describe above, the acquisition of FtsZ must be later than the primary endosymbiosis. MinC, a prerequisite for bacterial cell division, has not been identified in plants and algae. In early studies, chloroplast was found to contain a prokaryotic ribosome with the size of 70S (this is an estimate of the size based on sedimentation rate), which is sensitive to antibiotics, such as chloramphenicol and lincomycin, that inhibit prokaryotic ribosomes. Studies on the chloroplast ribosomes revealed, however, a number of chloroplast-specific components. The three-dimensional structure of chloroplast ribosome was elucidated by cryo-electron microscopy, and the presence of these plant-specific components was confirmed (Ahmed et al. 2016).

A core (consisting of α, β, β', and β'' subunits) of the prokaryotic RNA polymerase is encoded by the chloroplast genome, and is involved in the transcription of chloroplast genome. Nevertheless, the sigma subunit, a component that confers the promoter specificity to the core enzyme to form a holo enzyme, is encoded by the nuclear genome in multiple copies. The sigma subunit is translated in the cytosol, and then imported to the chloroplasts. Different isozymes of sigma are expressed in different tissues under different environmental conditions (for a concise review, see Börner et al. 2015). An additional RNA polymerase called RpoT or NEP (for nuclear-encoded polymerase) exists in land plants. It is encoded by the nuclear genome, but is likely to be acquired from a virus as a mitochondrial transcription enzyme in an early eukaryote. After gene duplication, a new form was targeted to chloroplasts in an early evolution of land plants (Sato 2001, 2002J). In angiosperms, the plastids present in the seeds are proplastids, a reduced form of plastids, which have neither activity of transcription and translation nor photosynthetic activity. Because of this, RNA polymerase encoded by the plastid genome cannot be synthesized in the proplastids. To start development of chloroplasts, an exogenous component must be provided to the plastid. RpoT turns on the switch of chloroplast development. The chloroplasts in angiosperms are, therefore, quite different from cyanobacterial cells, and are subjected to complex nuclear regulation depending on plant development.

Photosynthetic function is basically similar in cyanobacteria and chloroplasts. Both have Photosystems I and II and produce molecular oxygen. However, considerable differences are found in the composition of proteins that form oxygen-evolving complex: Cyanobacteria and red algal chloroplasts have PsbO, PsbU, and PsbV as components of the complex, whereas PsbO, PsbP, and PsbQ build up the oxygen-evolving complex of green algae and plants. The proteins with different names are indeed different in structure, and are not homologous. Interestingly, the red algal chloroplasts are also similar to cyanobacteria in the presence of phycobilisomes. In this case, we have to locate a discontinuous line between red algae and green algae.

Now we recognize that the conventional idea on the continuity from cyanobacteria to all plastids faces a serious problem that there are various types of discontinuities between them. Apparently similar characters can be supported by different kinds of proteins or genes. We consider these differences as a representation of discontinuities. One of the discontinuities occurred at the branch point of red and green algae after the formation of plastids. Another discontinuity in the use of mitochondrial RpoT in plastids should be placed during the evolution of green plants. Even the characters shared by all plastids could be acquired before or after the formation of plastids. Although we still need more data for the discussion on this point, we will try to examine available evidence in the next chapter. We will focus on whether some characters had been acquired by the host before the "primary endosymbiosis" (acquisition of rRNA).

Chapter 8
Re-examination of the "Endosymbiotic Event"

Endosymbiotic theory of organellar origin is an explanation of evolution. Visually understandable schemes showing cell engulfment events give misunderstanding that these are cytological events. Evolutionary explanation of various photosynthetic organisms necessitates variation in characters, inheritance of characters, and natural selection of various characters. The trait "photosynthesis in the chloroplast" depends on the characters encoded by the chloroplast genome as well as the characters encoded by the nuclear genome. If we admit that the chloroplast is an endosymbiont, chloroplast DNA and the characters encoded thereby are inherited with the chloroplast. However, proteins and lipids are not inherited by themselves.

When I learned the endosymbiotic theory of organellar origin in university courses more than 40 years ago, I felt that something remained to be explained. If a plant cell divides to engender daughter cells, then most of the proteins and lipids present in the original cell should be supplemented. If the proteins are encoded by the endosymbiont genome and expressed within the endosymbiont, then we can say that they will be "inherited" to the next generation. In the same way, if the biosynthetic enzymes of the lipids are encoded by the endosymbiont genome and expressed within the endosymbiont, then we can say that the lipids are "inherited" to the next generation (or more exactly, the same set of lipids are prepared in the next generation as if they are inherited). This seems to be valid in the chromatophores of *Paulinella* (Fig. 6.25 in Chap. 6), which synthesize their components almost autonomously, although we still need experimental evidence.

In the chloroplasts of present-day plants and algae, biosynthesis of membrane lipids is catalyzed by the enzymes encoded by the nuclear genome, and therefore the chloroplasts are not autonomous. If the enzymes are products of endosymbiotic gene transfer (EGT), namely, if the genes encoding the enzymes have been transferred from the endosymbiont to the nucleus, we can affirm that the endosymbiont deposited its genes to the host as a reservoir (Fig. 7.32 in Chap. 7). We know, however, that many of the enzymes involved in the synthesis of chloroplast lipids did not originate from cyanobacteria, but from various other bacteria, and that the remaining enzymes originated from the eukaryotic host. We cannot say that the lipid

N. Sato, *Endosymbiotic Theories of Organelles Revisited*,
https://doi.org/10.1007/978-981-15-1161-5_8

biosynthesis enzymes are characters inherited within the endosymbiont. The same is true for various proteins involved in other functions.

I ask the readers to think about this issue with an open mind. The notion that acquisition of an endosymbiont brings about a new character can be interesting as an idea, but it presupposes that the endosymbiont characters are inherited. Many researchers believe that the initial endosymbiont was autonomous like the chromatophores of *Paulinella* but that the autonomy became reduced with time by the replacement of genes determining the characters of the endosymbiont. We have to recognize, however, that the actual chloroplasts are strongly dependent on the host. We are not sure if the chromatophores will be more similar to the plant chloroplasts in the future. Likewise, it is not clear if the ancient chloroplasts were similar to chromatophores. In this chapter, we will examine the concept "endosymbiosis" in detail. A key notion is: "membranes are not inherited by themselves."

8.1 Possible Pathways of Formation of Plastids

We are now approaching the central problem in the endosymbiotic theory of organellar origin: what kind of evidence can unambiguously demonstrate endosymbiotic origin of organelles? Here we are discussing on plastids rather than mitochondria, because more solid evidence appears to be available for plastids. Then, what kind of evidence do we have to conclude that plastids originate from an endosymbiotic cyanobacterium? Two ways of thinking exist: cytological and phylogenetic ones. In cytological thinking, we would ask what kind of cytological changes could generate a plastid. In phylogenetic thinking, we would compare extant plastids and cyanobacteria. Existence of two ways of thinking is common in biology: Morange (2016, p. 261) showed two contrasting attitudes of embryonic development: evolutionary explanation by the recapitulation theory of Haeckel and mechanistic explanation of Roux. We are facing a similar situation.

To imagine possible pathways of plastid formation, we have a major premise that plastids are formed only from plastids. This was established by Schimper (1883). Likewise, cyanobacteria proliferate only by cell division. At some time in the history of the Earth, plastids came to exist. Is it possible that plastids formed spontaneously within a eukaryotic cell? This must be difficult because new genes are needed to form plastid structure. Figure 8.41 shows cytologically probable scenarios. Scenario C illustrates the commonly accepted theory of chloroplast formation by cyanobacterial endosymbiosis. But, as already described, we have to imagine a more complicated situation. In Scenario A, an ancient eukaryotic cell, already photosynthetic without a plastid-like structure, was subject to a metamorphosis to yield a plastid within the cell. In Scenario B, a non-photosynthetic eukaryotic cell, ready to make a plastid-like structure or already equipped with a plastid-like membrane structure, acquired necessary components to become a photosynthetic cell having a plastid. In the first scenario, a question remained as to whether photosynthesis was possible without specialized membranes. This can be imagined as a situation in

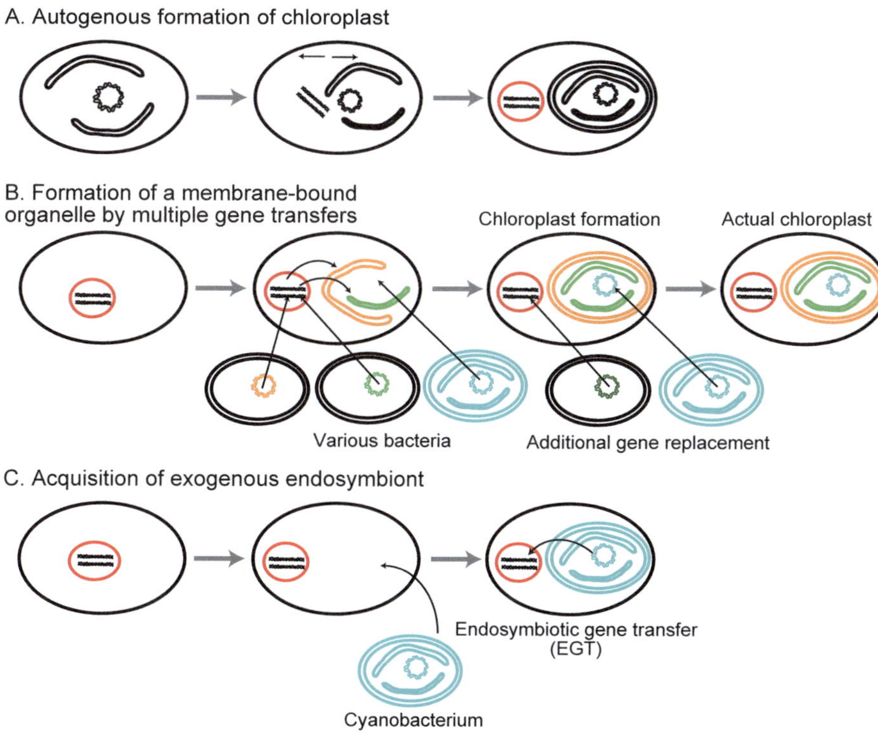

Fig. 8.41 Possible scenarios of chloroplast formation during the evolution of photosynthetic eukaryotes. Panel **(a)** is an autogenous (continuous) hypothesis. Panel **(c)** is the commonly accepted endosymbiotic theory. We will construct host-directed hypothesis based on Scenario (**b**). See text for details

which a cyanobacterial cell acquired a eukaryotic nucleus. In Scenario B, photosynthesis was not possible in the intermediate stages in which necessary components were acquired one by one. All intermediate cells do not have an advantage in natural selection, and this process is unlikely as an authentic evolutionary history. In addition, this cannot explain that the mechanism of photosynthesis is essentially similar in plastids and cyanobacteria.

Scenario A is similar to the autogenous, or continuous theory in the 1970s. This was invented to account for the presence of eukaryotic and prokaryotic cells, both having a similar mechanism of photosynthesis with oxygen evolution. Various authors (Uzzel and Spolsky 1974; Cavalier-Smith 1975; Bogorad 1975) tried to explain formation of an algal cell by membrane invagination or folding of internal membranes that led to the formation of a cell nucleus and other organelles within a cyanobacterial cell. This was logically possible without mitochondria. To explain also the origin of mitochondria, it was difficult to invent a plausible scenario. This scenario was not capable of explaining the phylogenetic difference of nuclear genome and plastid DNA. In the 1970s, Scenario C was the only alternative hypothesis that was capable of explaining the origins of both chloroplasts and mitochondria (see Fig. 1.9), although mitochondrial origin is not explicitly shown in Fig. 8.41c. This was also a plausible explanation, because it assumes a single emergence of photosynthesis, consistent with the principle of parsimony believed to dominate the Nature.

Scenario B, which was not considered in the 1970s, assumes another emergence of photosynthesis, which might not be plausible at a glance. We should suppose that many genes were introduced one by one from existing cyanobacteria. In this case, the acquired genes have different origins within the cyanobacteria, which was indeed demonstrated by phylogenetic analysis (Sato 2019). The problem of low selective advantages in the intermediate stages must be solved. A modified version of this scenario involves simultaneous acquisition of many genes at once. Nevertheless, Prof. Mitsuhiro Itaya (Keio University) who skillfully joined the genomes of a cyanobacterium and a bacillus did not obtain a photosynthetic bacillus. In this way, a mass of exogenous genes cannot allow instantaneous photosynthesis. We will come back to this issue later.

An endosymbiotic theory (Scenario C) was thought to solve various problems of the time (1970s) in a simple way. An entire cyanobacterial cell rather than a mass of cyanobacterial genes was incorporated within the cytosol of a eukaryotic cell to form an integrated, photosynthetically active, eukaryotic cell, by keeping the photosynthetic activity with various adjustments. In a popular explanation, the two membranes of cyanobacteria were covered by a phagosome membrane in the first step, and then one of the membranes disappeared to form the two envelope membranes of plastid (Fig. 1.9 in Introduction). As described earlier, the mechanism of membrane lipid synthesis is entirely different in cyanobacteria and plastids. The eukaryotic mechanism of lipid synthesis must be acquired at some time before or after the endosymbiosis. We have to integrate many more steps in the original, simple scenario to account for the properties of actual chloroplasts.

A good model for an endosymbiotic origin of plastids is the chromatophores of *Paulinella*, which were estimated to be acquired by an endosymbiotic event about 60-100 Mya (million years ago). The chromatophores retain the cyanobacterial system of lipid biosynthesis. We still do not know the DNA polymerase of the chromatophores. Researchers tend to believe that the chromatophores represent an intermediate stage of plastid evolution. Adjustment of regulatory or genetic systems in the endosymbiont and the host will change the endosymbiont into organelle during the evolution. Unfortunately, imagination of researchers is severely restricted by the visual image of endosymbiotic origin of organelles. They try to apply a smallest modification to the main scenario to keep it consistent with newly obtained experimental data.

8.2 Possible Stages of Endosymbiosis

Endosymbiosis is not a simple notion. It includes various different types of events, depending on researchers at different times with different interests. Mereschkowsky, Wallin, and Margulis believed literally that a chloroplast was a cyanobacterial cell living within the host cell. Until the middle of the twentieth century, the term "(endo) symbiosis" was used in the literal meaning of symbiosis between two organisms. In contrast, when we use the term "endosymbiosis" as an explanation of the origin of chloroplasts, chloroplasts are not independently living organisms. Chloroplasts depend on the nuclear genome. We have to classify various situations in which endosymbiosis is mentioned.

8.2.1 Independent Symbiosis (Parasitism)

This is a situation in which cyanobacterial cells proliferate within a host cell as independent organisms. The endosymbionts can be isolated and cultured outside the host cell. This was an example of the best ways of demonstration of endosymbiosis that Mereschkowsky and Margulis imagined. Until the 1970s, the cyanelles of *Paulinella* (now its cyanelles are called chromatophores) and glaucophytes were believed to be at this stage of endosymbiosis. The reversible symbiosis involving *Anabaena azollae* cells within the specialized organ of *Azolla* (Fig. 1.3 in Introduction) is an extracellular symbiosis. The spheroid body, an endosymbiont of cyanobacterial origin, within the cells of diatoms and haptophytes cannot be grown in isolation. The zooxanthelles (*Symbiodinium*) of corals are reversible endosymbionts, but they are eukaryotic cells. Currently, no cyanobacteria are known to be "reversible" endosymbionts.

8.2.2 Metabolic Dependency

This is a situation in which a metabolic relationship is established between the host and the endosymbiont. One partner provides nutrients to another. This can happen when one of the enzymes catalyzing an identical biochemical reaction shared by the host and the endosymbiont is lost by mutation in one partner. Then this partner becomes dependent on the other partner for the supply of the product of the reaction. Mutation is not an absolute requisite. Dependence on metabolic exchange is sufficient for this endosymbiosis. Metabolic exchanges are indeed known in various cases of endosymbiosis or symbiosis. In green paramecia, maltose is provided by the endosymbionts to the host. In root nodules, nitrogenous compounds such as amino acids or ammonia are provided by the endosymbiotic bacteria to the plant tissue, whereas carbon source is supplied by the plant tissue to the bacteria. The endosymbiotic bacteria (*Buchnera*) in the bacteriocyte of an aphid synthesize and provide necessary amino acids that cannot be synthesized by the host. *Buchnera* cells, on the other hand, lost most of the carbon metabolic pathways. Nodulating bacteria are independent organisms that can grow outside the nodules, whereas *Buchnera* cells are not viable outside the aphid. There are different degrees of metabolic dependency.

The chromatophores of *Paulinella* retain a large genome (encoding about 1000 genes) of the cyanobacterial origin, and have the ability to synthesize fatty acids and lipids. Although they lost some genes (such as the *murF* gene encoding an enzyme for a step in peptidoglycan synthesis: note that a nuclear *murF* gene of a bacterial origin exists), they are in a situation quite similar to metabolic dependency.

8.2.3 Complete Dependency by Endosymbiotic Gene Transfer

In the current understanding of chloroplast “endosymbiosis,” many genes originally encoded by the endosymbiont genome were transferred to the cell nucleus by an unknown mechanism (this is called endosymbiotic gene transfer or EGT: see Fig. 1.9 in Introduction), and subsequently lost from the endosymbiont genome. Accordingly, the endosymbiont cannot grow outside the host cell. A question arises as to the mechanism of gene expression of the transferred genes: Namely, the expression mechanism of the endosymbiont genes is prokaryotic, and is not compatible with the mechanisms of gene expression within the cell nucleus. If the genes transferred from the endosymbiont are eventually expressed in the nucleus, the proteins synthesized on the cytoplasmic ribosomes must be transported into the endosymbionts (chloroplasts) to function. Nuclear expression requires a eukaryotic promoter. According to Prof. Jun-ichi Obokata (Kyoto Prefectural University), cryptic promoters exist in the nuclear genome to account for eventual expression of a transferred gene. In contrast, chloroplast import of polypeptides is mediated by the translocon (protein translocating machinery) located in the envelope membrane. Every polypeptide to be imported must bear an *N*-terminal transit sequence, which is different in different polypeptides. Probably, all transit peptides were acquired independently. Comparative genomic studies suggested that about several hundreds or a thousand genes were transferred from the endosymbiont to the nucleus (Martin et al. 2002; Sato et al. 2005).

8.2.4 Symbiotic Relationship Dependent on Horizontally Transferred Genes

The gene transfers in the previous situation are easy to understand, but another, somewhat mysterious type of gene transfers involves nuclear genes that were supposed to originate from various bacteria other than cyanobacteria, supposedly legitimate origin of plastids. Results of genomic analyses pointed out that non-cyanobacterial genes are abundant in plant and algal nuclear genomes (Qiu et al. 2013; Ku et al. 2015). As described in Chap. 7, many enzymes involved in the synthesis of glycolipids and peptidoglycan, which had been regarded as good evidence for the cyanobacterial origin of chloroplasts, turned out to originate from various bacteria other than cyanobacteria. Use of homologous genes in phylogenetic analysis restricted genomic analysis to detection of homologous genes. In contrast, the pathways of lipid biosynthesis and tetrapyrrole synthesis include certain reactions, in which nonhomologous enzymes are involved, or a particular step, in which entirely different reactions are used. These are discontinuities that are not detected by homology-based genomic analysis.

I have to recall protoporphyrinogen IX oxidase in tetrapyrrole (heme and chlorophyll) synthesis (Kobayashi et al. 2014), listed as a discontinuity in Table 7.12. Three nonhomologous, isofunctional enzymes, HemY, HemJ, and HemG, act as protoporphyrinogen IX oxidase. The three enzymes are found even in related species. Many horizontal gene transfers (in which genes of unrelated organisms are incorporated and become functional for an unknown reason, such as direct infection, mediation by a virus, and DNA transfection) could account for the complexity in the phylogenetic distribution of the three enzymes, namely, any gene encoding one of the isofunctional enzymes was replaced by another. A very curious case was found in *Prochlorococcus* (cyanobacteria), in which *hemJ* and *hemG* are placed at exactly identical site (with respect to the flanking genes) within the genome in closely related species. This perfect replacement is comparable to a result of genetic engineering by a well-trained researcher (see Kobayashi et al. 2014 for details). This example gives us an important lesson that gene origin should be traced in individual reactions, rather than identified by statistical analysis at the whole genome level.

When were the exogenous genes acquired? After the acquisition of the endosymbiont? Could the exogenous genes be acquired before the endosymbiont? If the exogenous genes to be used after the endosymbiosis had been acquired before the endosymbiosis, they did not confer selective advantage to the host, and would be lost rapidly. In evolutionary biology, interesting notions, preadaptation or exaptation, are used to explain various evolutionary changes. In exaptation, a body part having a certain function will gain a new function, which triggers a remarkable evolutionary change. The body part may be microscopic or macroscopic. Bird feathers are known as a typical example of exaptation. In the origin, the feathers of reptiles functioned in keeping body temperature. They were reutilized for flying in birds.

In mitochondria, Pittis and Gabaldón (2016) showed that the genes originated from α-proteobacteria were acquired later than the genes from other bacteria. This was evidence for the "mito-late" hypothesis, in which an ancestral eukaryotic cell had acquired many genes from various bacteria by horizontal gene transfer before mitochondrial formation, and then these genes became functional within the mitochondria. Another study (Harish and Kurland 2017) showed that 97% of the 615 functional domains of all mitochondrial proteins originated from the universal common ancestor (UCA) but not particularly from α-proteobacteria. Because we can easily imagine that most of the mitochondrial proteins had been host proteins that changed intracellular localization, we should still be prudent to say that the origins of mitochondrial proteins can be estimated by such global, statistical analysis. Nevertheless, we are convinced that a number of proteins that have existed since long before the mitochondrial endosymbiosis are currently functioning within the mitochondria.

I already used a similar method to estimate the time of acquisition of chloroplast enzymes involved in chloroplast lipid biosynthesis (Table 7.11, Chap. 7). The results showed that some proteins with small values of relative stem lengths are likely to be acquired after the primary endosymbiosis as defined by the acquisition of rRNA. Some other proteins with larger relative stem lengths were supposed to be acquired before the primary endosymbiosis. The remaining proteins could be acquired with the primary endosymbiosis or at times close to it.

We found that the glycolipid synthesis enzymes, SQD2, SQD1, and MGD1, were acquired by the host before the primary endosymbiosis (acquisition of rRNA). Glycolipids do not contain phosphorus, in contrast with phospholipids that are common constituents of cell membranes. If phospholipids can be replaced with glycolipids, then cellular requirement of phosphorus can be reduced. Phosphorus is a minor element in the Earth crust (11th most abundant, 0.1% in the crust), but very poor in ocean water ($6–8 \times 10^{-6}\%$) because of low solubility of phosphate salts, and severely limits survival of organisms especially in aquatic environments (Data from bionumbers web site: BNIDs 110,362 and 107,256). Sulfur in SQDG is also a minor element, but more available in water (17th abundant, 0.03% in the crust, 0.1% in ocean water). Phosphorus is a constituent of nucleic acids and high-energy compounds such as ATP. Requirement of phosphorus in cell metabolism is much larger than that of sulfur. Chloroplasts and cyanobacteria contain multilayered thylakoid membranes to efficiently harvest light for photosynthesis. If thylakoid membranes were to be composed entirely of phospholipids, then cellular requirement of phosphorus would be very high. Use of glycolipid in thylakoid membranes allows the photosynthetic organisms to survive at a lower concentration of phosphate in the environment. This is a good reason to assume that glycolipid biosynthesis capability (acquired from various bacteria) had a selective value even before the formation of chloroplasts.

8.2.5 Massive Horizontal Gene Transfers

This last situation that might look unrealistic assumes that many genes were acquired at a time by horizontal gene transfer. Lane (2017) described a similar scenario for the origin of eukaryotes. If all necessary genes are introduced as a mass into the host cell from cyanobacteria and other bacteria, then chloroplasts might show up. As shown in Table 7.11 (Chap. 7), various non-cyanobacterial proteins, not just chlamydial proteins, were acquired at the time of the primary endosymbiosis. We may also understand this as massive horizontal gene transfers, without using the term "primary endosymbiosis." If we do not consider structural entity, this is no longer an endosymbiosis.

8.3 Fatty Acid Hypothesis on the Origin of Photosynthetic Eukaryotes

Now I ask the readers to consider three points related to the endosymbiotic theory of organellar origin: They are (1) fatty acid hypothesis, (2) host-directed plastid formation, and (3) multiple origins of plastids. These three are not alternative hypotheses on an identical problem, but independent hypotheses that explain different but related aspects of plastid evolution. I will explain them in three sections beginning from this one.

8.3.1 Fatty Acid Synthases in Animals and Fungi

What is a eukaryotic cell? What is a plant? We now come back to these fundamental questions. Then, we find a curious contrast between plants and animals. As described previously, plants possess their own fatty acid synthesis system, possibly originating from cyanobacteria, in their plastids, and synthesize fatty acids by themselves. High productivity of fatty acids allows various oil crops to accumulate large amounts of oil in their seeds.

Animals, on the other hand, possess cytosolic fatty acid synthase and synthesize fatty acids. This enzyme is, in contrast with the corresponding enzyme system in plants and bacteria, a large single, multifunctional polypeptide, integrating various enzyme activities. In yeasts, fatty acid synthase consists of two multifunctional polypeptides. Some fungi have fatty acid synthase with a large single polypeptide. Significant differences in the arrangement of functional domains suggest that the fatty acid synthases of fungi and animals are different in origins (Maier et al. 2010). The studies on these enzymes began in the 1960s, but were retarded by their very large sizes. Structural analysis and molecular analysis of the fatty acid synthases advanced only in these two decades. Interestingly, phylogenetic analysis indicated that these enzymes originated from bacterial polyketide synthases (Jenke-Kodama et al. 2005). Polyketides are various functional compounds produced by microorganisms from acetyl CoA as a precursor. They include many useful antibiotics. A single enzyme catalyzes the synthesis of a large, complex polyketide molecule by adding various simple building blocks in an ordered sequence. Fatty acid synthases of animals (clade IV), and fungal single polypeptide fatty acid synthases (clade VII) are close to the clade VII polyketide synthases of bacteria, whereas yeast fatty acid synthases (clade III) are related to clade II polyketide synthases of actinobacteria, such as *Corynebacterium* and *Mycobacterium*. It is now clear that the animal and fungal fatty acid synthases are not related to the fatty acid synthases of plants and cyanobacteria.

8.3.2 Were Ancestral Eukaryotes Able to Synthesize Fatty Acids?

A recent study reported that arbuscular mycorrhizal fungi do not synthesize fatty acids, but that their life entirely depends on the supply of fatty acids by the host plants (Luginbuehl et al. 2017). This might seem an exceptional case due to symbiotic relationship, but we can also imagine that this is a default state of fungi in general, although the mycorrhizal fungi might have lost fatty acid synthase secondarily. Land plants appeared on the Earth later than fungi. The ancestors of mycorrhizal fungi had to obtain fatty acids from organisms other than land plants if they were unable to synthesize fatty acids. Animals do not need to synthesize fatty acids by themselves either, if they eat plants and bacteria. Humans indeed depend on necessary fatty acids such as linoleic and linolenic acids from dietary nutrients. Animals perhaps have to synthesize saturated fatty acids by themselves because plant lipids contain only a low level of saturated fatty acids. A good example of saturated fatty acids in animals is dipalmitoyl phosphatidylcholine that acts as a surfactant on the surface of pulmonary alveoli. Animal fats are rich in saturated fatty acids, such as lard and tallow, and therefore are solid at ambient temperatures. Ancestors of animals must have acquired microbial fatty acid synthase closely related to polyketide synthase to synthesize saturated fatty acids that are poor in dietary vegetables. Note that animals and fungi belong to a lineage different from plants (Fig. 6.20, Chap. 6). We have to understand that plants and algae are quite different from animals and fungi in phylogeny and metabolism. Biochemists sometimes used an expression "from yeast to human" to generalize their finding, but this refers to only one group of eukaryotes, namely, Opisthokonta (now Amorphea). Organisms in this group possess high-molecular-weight fatty acid synthases, which are fundamentally different from the fatty acid synthases of plants and algae. Some extant fungi, as I explained, do not even have any fatty acid synthase. Under these circumstances, it is a reasonable assumption that ancestral eukaryotes did not synthesize fatty acids.

Many scientists consider that Archaea was an origin of eukaryotes. Recent studies suggested that the Superphylum TACK (named after the initials of the four member groups of the superphylum) in Archaea is very close to eukaryotes. Especially, Lokiarchaeota is supposed to be the origin of eukaryotes (Fig. 8.42) (Spang et al. 2015). An important characteristic of Archaea is the presence of ether lipids (Fig. 8.43) consisting of isoprenoids (polymer of branched C5 units), rather than diacylglycerol-type lipids, such as phosphatidylethanolamine, present in bacteria and eukaryotes. Eukaryotes and some bacteria also contain sphingolipids, derivatives of sphingosine bases, which are synthesized by condensation of palmitic acid and serine. Therefore, synthesis of fatty acids is a prerequisite for sphingolipid synthesis.

A minute, but essential difference in the lipids of Archaea and other groups of organisms is the stereochemical configuration of the glycerol moiety of lipids. In Archaea, glycerol 1-phosphate is the backbone of lipids, whereas, in bacteria and eukaryotes, glycerol 3-phosphate is used as the carrier of acyl groups. The two glycerol phosphates are optical isomers, and are not convertible to each other without cleavage of a covalent bond. Hypotheses on the origin of bacteria and Archaea must answer this question. The last universal common ancestor (LUCA) of both bacteria and Archaea was supposed to contain both types of lipids, only one of which was used in each lineage after separation (Caforio and Driessen 2016). A major critique of this kind of hypothesis is that it forces us to assume everything in the ancestral organism.

As to the origin of eukaryotes, I suppose that ancestral eukaryotes were not able to synthesize fatty acids. This is exactly the property of Archaea! We are not sure how the first eukaryote switched from ether lipid to diacyl lipid. As bacteria already synthesized fatty acids at the time, the first eukaryote was able to use them. Evolution from Archaea to eukaryote involved a change in lipid type, which necessitated ancestral eukaryotes to ingest bacteria. Most of the present-day Archaea live in extreme environments, such as alkaline, hot, or cold habitats, to which no other organisms can access. Under these conditions, use of chemically stable ether lipids is advantageous. Currently, archaeal cells might not be able to compete efficiently

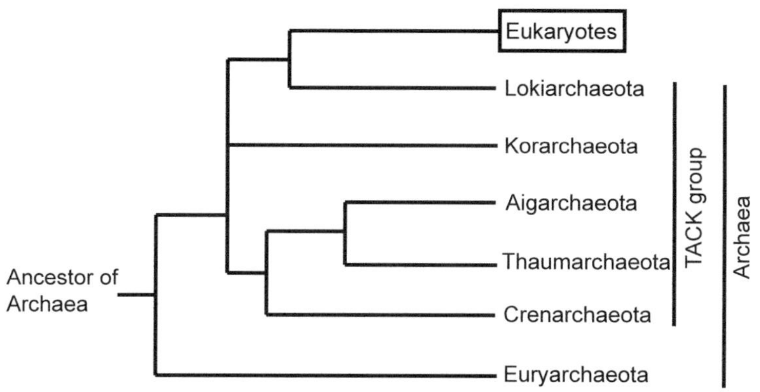

Fig. 8.42 Relationship of eukaryotes with Archaea. Redrawn as an original figure after Spang et al. (2015)

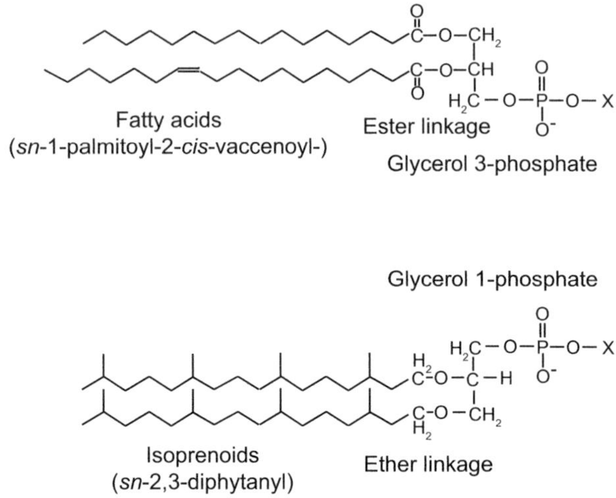

Fig. 8.43 Lipids of bacteria and archaea. Acyl groups and isoprenoid groups are typical examples. See text for details. Original figure

with eukaryotes and bacteria in other habitats. A hypothetical, extinct ancestor of eukaryotes, having changed its membrane lipid composition, might not be able to survive without enough bacterial baits. At that time, they had two choices: either to acquire fatty acid synthase related to polyketide synthase from bacteria, or to acquire cyanobacterial system of fatty acid synthesis together with photosynthesis.

The research group of Prof. Bill Martin in Düsseldorf, Germany, has been studying the origin of eukaryotes by endosymbiosis of mitochondria into an archaean cell (Martin et al. 2015). Gould et al. (2016) from the group proposed the vesicle secretion hypothesis, in which the plasma membrane of archaean cell was replaced by a membrane produced by mitochondria and the endomembranes of eukaryotic cell originated from mitochondrial membranes. This is a hybrid of endosymbiotic theory and autogenous theory. In this case, where has the fatty acid synthase gone, which must have been present in the α-proteobacterial endosymbiont? It is true that present-day mitochondria retain a condensing enzyme and an acyl-carrier-protein (ACP) (Fig. 7.35, Chap. 7). Nevertheless, these enzymes are not capable of synthesizing all fatty acids required by the entire cell. The power of mitochondria is not sufficient for the task. Another problem related to the vesicle secretion hypothesis is, as already pointed out in Sect. 6.5.4, the difference in enzyme catalyzing the lipids between bacteria and mitochondria.

8.3.3 *Photosynthesis-Driven Fatty Acid Synthesis*

Let us think about our foodstuffs here. We take various oily foods, such as snacks, fried chicken, and beefsteak, in daily life, but we cannot easily imagine how difficult it is to produce oil. Why do we get fat if we take too much oil? Because oil is rich in calorie: the calorie value of oil is 9 kilocalories per gram, whereas that of carbohydrate is only 4 kilocalories per gram. In other words, oil is a high-calorie food. It contains much free energy in a condensed form. Chemically, fatty acids that compose oil contain about two hydrogen atoms per carbon atom, or the oxidation number of carbon is minus 2. In contrast, the oxidation number of carbon dioxide is plus 4, and that of carbohydrate is zero. To decrease the oxidation number, a reducing power (or hydrogen) is necessary. In photosynthesis, carbon dioxide is reduced to form carbohydrate, with a decrease in oxidation number by 4. To accumulate oil in oil seeds, the oxidation number must be further decreased by 2. This is accomplished by the reducing power of photosynthesis. Evidently, oil is a product of photosynthesis. In chloroplasts, photochemical reaction is performed in multilayered thylakoid membranes composed of lipids, which are products of photosynthesis. This indicates that a part of the reducing power obtained by photosynthesis is reinvested for constructing photosynthetic machinery, thus enabling expanded reproduction. Non-photosynthetic organisms use carbohydrates produced by photosynthesis in plants and algae. They degrade a significant part of the food carbohydrate to obtain free energy to synthesize lipids from the remaining carbohydrate for the construction of their cell architecture. Although mycorrhizal fungi obtain directly fatty acids from host plants as described above, all non-photosynthetic organisms utilize indirectly the carbohydrates and lipids that photosynthetic organisms produce.

Based on the importance of lipids, we have a serious question about how the initial eukaryotic cell that just acquired mitochondria (as imagined by Margulis) could manage to obtain enough fatty acids. Contemporary animals and fungi can obtain fatty acids by eating (or digesting by parasitism) plants and algae. However, before the creation of chloroplasts, there were only primitive eukaryotes and bacteria. Bacteria were able to synthesize fatty acids by themselves, but primitive eukaryotes must prey bacteria to obtain fatty acids. Mitochondria have insufficient activity of fatty acid synthesis, because mitochondria are catabolic organelles. It is true that NADH (a reductant) can be produced in non-photosynthetic organisms from carbohydrates by the central carbon metabolism including glycolysis and citric acid cycle, but most of it must be consumed for ATP production. The amount of NADH (or NADPH) that can be used for fatty acid synthesis is quite limited. The supply of carbohydrates must also be limited for primitive eukaryotes that fed on bacteria or other primitive eukaryotes. In the primitive Earth, reducing power was only provided by ammonia and hydrogen sulfide (and perhaps hydrogen gas) released from the Earth crust. All organisms must have been relying on this limited reducing power to produce carbohydrates or other organic substances.

In contrast, photosynthesis provides a super reducing power, or unlimited, new reducing power. The establishment of chloroplasts in algae (and later plants) expands the photosynthetic capacity, previously limited to cyanobacteria, and enabled all life to proliferate on the planet. Chloroplasts were able to synthesize large amounts of fatty acids using the reducing power of photosynthesis. Then, non-photosynthetic organisms no longer had to synthesize fatty acids by themselves, and they preyed algae. Among all eukaryotic organisms existing today, only plants and algae possess prokaryotic fatty acid synthase. They acquired most enzymes involved in photosynthesis and fatty acid synthesis (but not lipid synthesis) from cyanobacteria or ancestors of cyanobacteria. The acquisition of these two activities as a set was meaningful, namely, fatty acid synthesis was an integral part of photosynthesis that enabled algal life. However, it remains to be studied why the condensing enzymes (KAS I/II) were acquired from Chlamydiae or green bacteria. We can ask how the enzymes of different origins work in collaboration. Alternatively, the origin of chloroplasts might involve these bacteria, not just cyanobacteria.

8.4 Host-Directed Plastid Formation

All these discussions led us to think about another hypothesis on the origin of plastids. It is now clear that the stereotyped illustration showing that a cyanobacterial cell, retaining its membrane systems, became a plastid is no longer valid. In the explanation of the secondary endosymbiosis, as illustrated by Cavalier-Smith and others, endosymbiont membranes persist during the evolutionary process. This is not correct.

We have to develop a hypothesis in which organellar membranes did not originate from the endosymbiont. I propose a hypothesis on the host-directed plastid formation as a derivative of the scheme in Fig. 8.41b. One of the characteristics of eukaryotic cell is the ability of autophagy, which became famous by the Nobel Prize given to Professor Yoshinori Ohsumi (Tokyo Institute of Technology and the University of Tokyo). Autophagy is present in many eukaryotes except red algae. An important property of autophagy is to create a membrane de novo that will cover an organelle of target (Liu and Bassham 2012; for Japanese introduction, see Yoshimoto 2014J; Ishida 2014J). According to the current scheme, a protein called ATG5 binds a phosphatidylethanolamine (PE) molecule, which serves as a basis of building a new membrane. PE is present in all eukaryotes (and many bacteria) and acts as a key compound in various cellular functions, but not just as a membrane component. The newly synthesized, autophagic membrane encapsulates the target cell structure to be degraded, and then fuses with the vacuolar membrane to throw the target into the vacuole for final digestion. An important property of eukaryotic cells, not shared by prokaryotic cells, is this ability to create a new membrane from scratch. However, as I explained earlier, the ancestor of eukaryotes was not able to synthesize fatty acids. I wonder how the autophagy system using a diacyl lipid PE originated in the ancestor of eukaryotes. Autophagy could be a recycling system for reuse of diacyl lipids.

Suppose that an ancestral eukaryotic cell (future photosynthetic algal cell) acquired genes for the synthesis of glycolipids from some bacteria (before the endosymbiosis). Glycolipids do not contain phosphorus, and therefore biomembranes with glycolipids are advantageous in phosphate-limited environment. Plants indeed accumulate a galactolipid DGDG (digalactosyl diacylglycerol) under phosphate limitation, and use it as a component in the plasma membrane as well as chloroplast membranes. If a eukaryotic (host) cell possessing a galactolipid biosynthetic activity captured a cyanobacterial cell by chance, the supply of galactolipids by the host could result in the loss of galactolipid biosynthetic genes in the cyanobacterial cell. In bacterial genetics, we sometimes encounter spontaneous inactivation and loss of the genes that are not necessary for the survival under a controlled condition. The same must be true for the hypothetical host-cyanobacterial system. The cyanobacterial cell that depended on the host for the supply of galactolipids were no longer able to live independently and became an absolute endosymbiont. This mechanism of dependency was possible with any necessary gene. If any gene necessary for the survival was lost and if its function was genetically complemented by the host, then the endosymbiont would not be able to escape from the host cell (Sect. 8.2.4). This must be a genetic trap invented by the host to capture a cyanobacterial cell.

An alternative scenario is that the host cell prepared a membrane composed of glycolipids by a mechanism like autophagy, and then captured a cyanobacterial cell within it. The original cyanobacterial membranes were degraded, and then photosynthetic systems encoded by the cyanobacterial genome became embedded within the autophagic membrane, and began functioning. This hypothesis is an extreme idea combining the autogenous theory and horizontal gene transfer. We might also be able to consider an intermediate hypothesis between the trap scenario and the autophagy scenario: A cyanobacterial cell, once entrapped by the host, became encapsulated by an autophagosome, but remained there without being sent to the vacuole. The original cyanobacterial membranes were degraded in some way, and replaced by the host membranes to keep the photosynthetic activity.

Until now, I did not talk about protein import into the chloroplasts. In the extant plant and algal cells, all plastid proteins encoded by the nuclear genome are translated in the cytosolic ribosomes, and then transported into the plastids through the protein-translocating machinery called translocon (see Paila et al. 2015 for a review). Some translocon components originate from cyanobacteria, whereas others are eukaryotic. In Matsuzaki et al. (2004), I showed that the translocon components are shared by the green lineage (green algae and land plants) and the red lineage, suggesting that the translocon system established early after the plastid formation was conserved after the separation of the green and the red lineages. This was verified later by including glaucophytes and other algae (Gould et al. 2008; Price et al. 2012). In short, this was evidence for the monophyly of green, red, and glaucophyte lineages, and supported the Archaeplastida notion. Translocon is indeed a mark that tells the identity of organelle. We can imagine a hypothetical experiment in which all translocon components are integrated into liposomes, which might then become organelles similar to chloroplasts if introduced into a plant cell. In this sense, the translocon that enables transport of many nuclear-encoded proteins into the plastid defines the identity of plastid. This, in turn, determines the nonautonomous nature of the plastid. From the viewpoint of the host-directed establishment of plastid, the host was supposed to prepare a translocon system to convert its intracellular membranes into plastid membranes.

8.5 Mosaic Nature of Plastid Genome: Multiple, Complex Origins of Plastids

We already found that many enzymes involved in the synthesis of chloroplast membrane lipids did not originate from cyanobacteria. The same is true for some other chloroplast enzymes. In Fig. 7.33 (Chap. 7), we also found two types of phylogenetic trees that showed close relationship between chloroplast enzymes and cyanobacterial homologs. Type 2 trees have not been recognized as distinct from Type 1 trees. In Type 1 trees, chloroplast enzymes diverged as a single clade from cyanobacterial homologs within the cyanobacterial diversity. Typical trees of this type are shown in Figs. 6.21, 6.22, and 6.23 (Chap. 6). Type 1c trees including only cyanobacterial and chloroplast enzymes also present this pattern of phylogeny, if an appropriate taxon (usually *Gloeobacter* protein) is taken as a basal group.

8.5.1 PsaA and PsaB

An example of Type 1c tree is shown in Fig. 8.44, a phylogenetic tree of PsaA and PsaB, in which a root is reasonably estimated between the clades of PsaA and PsaB. It should be noted, however, that the branch point of plastid clade was located at the base of *Synechococcus-Prochlorococcus* subclade. This position is clearly different from the deep origin of chloroplast rRNA (Fig. 6.23).

In Type 2 trees (Fig. 7.33), the cyanobacterial clade and the chloroplast clade are sister groups, namely, the two clades diverged from a common ancestor. I cannot imagine what this common ancestor was. Type 2 trees have been obtained in various phylogenetic analyses in the past, but were interpreted to show close relationship between cyanobacteria and chloroplasts. We will discuss this problem later.

8.5.2 Ribosomal Proteins

We have to be careful in analyzing phylogenetic trees of small proteins, which do not give sufficient phylogenetic signals to identify different clades. For example, many ribosomal proteins encoded by the chloroplast genome give Type 2 phylogenetic trees when analyzed individually. They are Rpl2, Rpl14, Rpl16, Rpl20, Rpl23, Rps8, Rps11, Rps12, Rps18, and Rps19. However, when all these sequences were concatenated as a single, long amino acid sequences, a Type 1 tree was obtained (Fig. 8.45). This is not magic. In short sequences, a single amino acid change has a heavy weight within the total sequence similarity, and can give an excessive signal for phylogenetic reconstruction. In a long, combined sequence, the weights of individual mutations are more balanced, and contribute equally in a more plausible phylogenetic tree. Note that a *Paulinella* sequence is found among the cyanobacterial

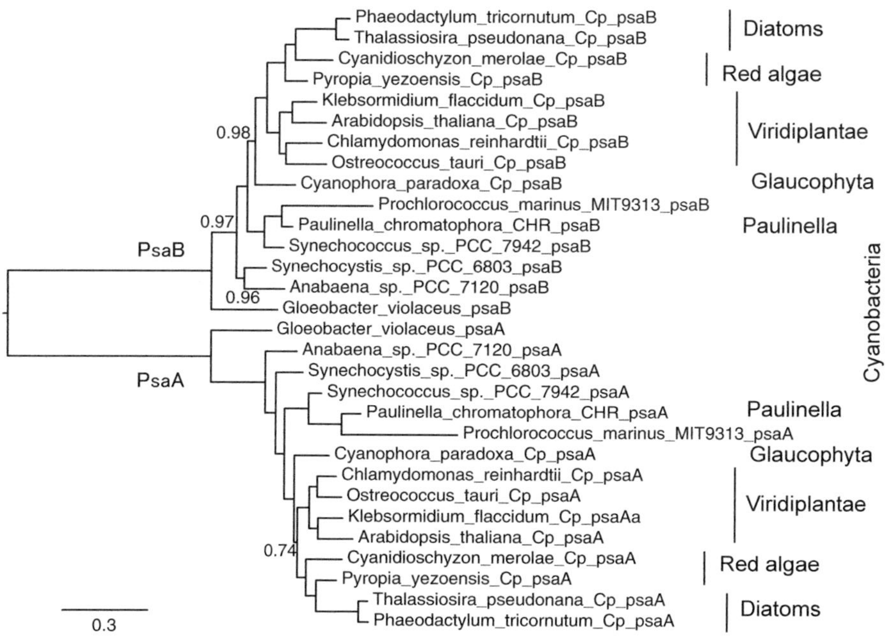

Fig. 8.44 Phylogenetic tree of PsaA and PsaB, two large components of the photosynthetic reaction center in Photosystem I. To determine the root in a phylogenetic tree of cyanobacterial and chloroplast proteins, bacterial orthologs are used if available. For photosynthetic proteins, this is not possible because no bacterial homologs are known. Nevertheless, rooting is possible for the two homologous proteins, PsaA and PsaB, which are structurally similar and considered to diverge from a common ancestor: a root is reasonably placed between the two clades in a combined tree of these two proteins. As a result, we can see that both proteins diverged first within cyanobacteria, then introduced into chloroplasts, and finally diversified within algae and plants. Diatom chloroplasts are considered to originate from chloroplasts of red algae. A close examination of the tree indicates some differences. We do not know if this is a technical problem or it reflects the real phylogenetic history of the proteins. This tree was obtained by the Bayesian Inference method. All branches are supported with 100% confidence unless otherwise indicated. Revised from the original Japanese version, and reproduced from Sato (2019) with permission of the publisher

sequences, suggesting a recent origin of its chromatophore. Here, we are focusing on the primary endosymbiosis of chloroplasts, keeping the chromatophore of *Paulinella* as a member of cyanobacteria.

8.5.3 *Ndh Proteins*

Another case of complexity is found with Ndh proteins that are components of the photosynthetic complex I of chloroplasts. The name Ndh indicated the similarity of these proteins to the subunits of mitochondrial NADH dehydrogenase. Currently,

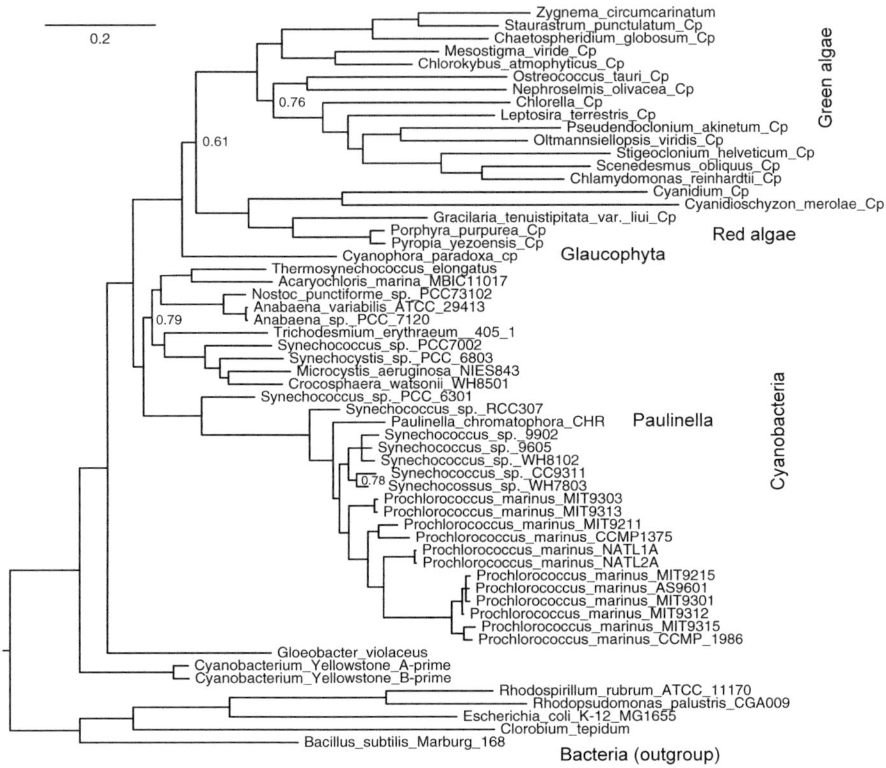

Fig. 8.45 Phylogenetic tree of concatenated sequences of ribosomal proteins. Type 2 phylogenetic trees are found for various ribosomal proteins, when analyzed individually. However, concatenated sequences of such unusual ribosomal proteins give a Type 1 tree as shown in rRNA (Fig. 6.23, Chap. 6), except the relative branching orders of *Gloeobacter* and the two Yellowstone strains. Probably, small sizes of ribosomal proteins made it difficult to reconstruct a reliable phylogenetic tree. The result shown in this figure is consistent with the idea that all ribosomal proteins and rRNA were introduced into chloroplasts at the same time. This tree was calculated by the Bayesian Inference method. All branches are supported at 100% or near 100% confidence levels, except indicated otherwise. Reproduced from Sato (2019) with permission of the publisher

however, the chloroplast complex lacks an NADH binding domain, and is considered as ferredoxin-dependent dehydrogenase involved in the removal of excessive reducing power in Photosystem I (Schuller et al. 2019; Laughlin et al. 2019). More than 20 subunits (including those that are not called Ndh*) are known in cyanobacteria. Eleven of them, NdhA, B, C, D, E, F, G, H, I, J, and K, are encoded by the chloroplast genomes of land plants and some green algae in Streptophyta (for a review, see Shikanai 2016). The chloroplast genome of *Nephroselmis* lacks *ndhJ* gene. I performed phylogenetic analysis using a limited number of species for several subunits. NdhD consists of many paralogs, and was not included in the phylogenetic analysis. Small subunits were not included, either. Sequences of land plants were not included to focus on the initial branching of the green lineage. Ndh B, C, E, G, and K give Type 2 trees or more strange trees when analyzed individually. If these five components are analyzed as a combined protein, we still obtained a Type 2 tree. But if all nine Ndh components (except D and J) are concatenated, a Type 1 tree was constructed (Fig. 8.46). The green algal clade diverged early after *Gloeobacter*. This branching is not identical to that found in rRNA (Fig. 6.23) or ribosomal proteins (Fig. 8.45). It is still not clear if the *ndh* genes were introduced into the plastid genome as a package with ribosomal components.

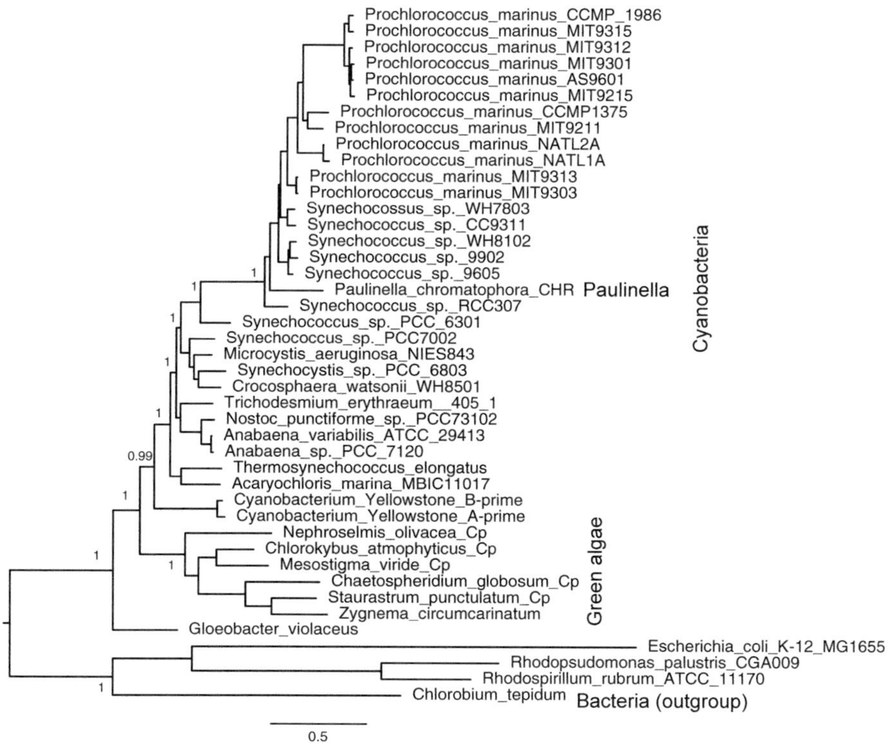

Fig. 8.46 Phylogenetic tree of NDH proteins. This is a phylogenetic tree of concatenated NdhA, B, C, E, F, G, H, I, and K. Type 2 trees are obtained when NdhC and E are analyzed individually. Individual trees of NdhG, H, and I are also strange in branching pattern. Concatenated sequences gives a Type 1 tree, but the origin of the green algal sequences diverged very early after *Gloeobacter*. It is still unclear whether chloroplast Ndh proteins originated from a single event of gene transfer from cyanobacteria. Reproduced from Sato (2019) with permission of the publisher

8.5.4 RubisCO

The carbon fixation enzyme RubisCO (ribulose 1,5-bisphosphate carboxylase/oxygenase) consists of two subunits, called large (L) and small (S) subunits. The subunit L is encoded by the plastid gene *rbcL* in all plants and algae. The subunit S is encoded by the plastid genome (gene name: *rbcS*) in red algae and some other algae, whereas it is encoded by the nuclear genome (*RbcS*) in green algae and land plants. Both *rbcL* and *rbcS* genes of red algae are known to originate from α-proteobacteria possibly by horizontal gene transfer (Tabita 1999; Sato 2006) as noted previously (Sect. 7.6). The situation is quite complicated as shown in Fig. 8.47. The RbcL of

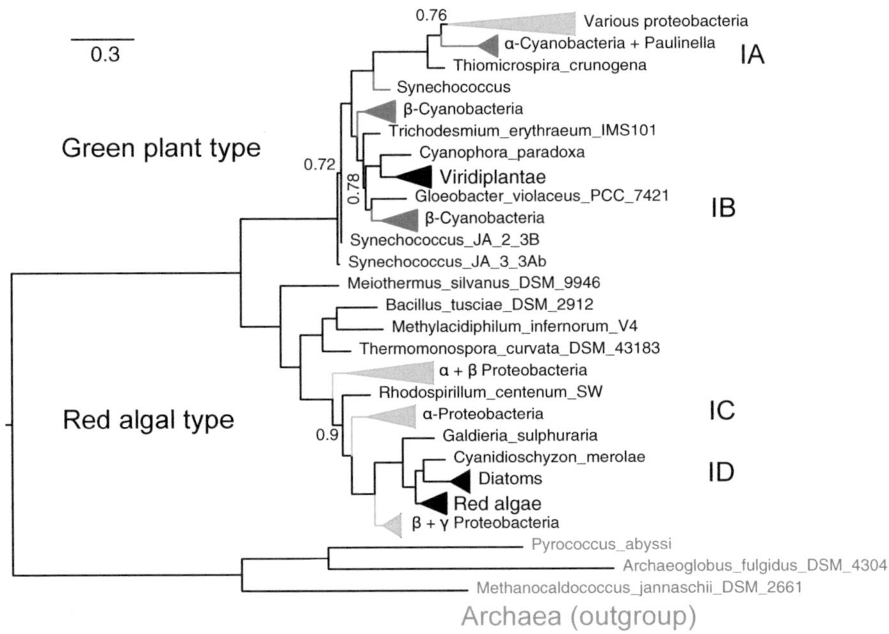

Fig. 8.47 Phylogenetic tree of RubisCO large subunit. RubisCO (Ribulose bisphosphate carboxylase/oxygenase) consists of two types of subunits, large and small, which are encoded by *rbcL* and *rbcS* genes, respectively. RbcS proteins are small and highly divergent, partly because green plant homologs are encoded by the nuclear genome while homologs in other algae are encoded by the chloroplast genome. That is why RbcS was not used for phylogenetic analysis. The *rbcL* gene is encoded by the chloroplast genome in all plants and algae. RbcL is often used to estimate phylogenetic relationship in plants. RubisCO is, in fact, present in a wide variety of organisms, including both photosynthetic and non-photosynthetic ones. Photosynthetic bacteria that do not produce oxygen also possess this carbon-fixation enzyme. The tree indicates that RbcL in the green lineage originated from cyanobacteria, whereas the counterpart in the red lineage from proteobacteria. The RbcL of α-cyanobacteria originated from other proteobacteria. This tree was calculated by the Maximum Likelihood method, and major branches are supported at 100% confidence level except otherwise specified. IA–ID indicate classification by Tabita (1999). Reproduced from Sato (2019) with permission of the publisher

α-cyanobacteria (*Prochlorococcus* and marine *Synechococcus*) is considered to originate from γ-proteobacteria.

The plastid genome is usually considered simply a heritage of cyanobacteria, but even the gene(s) of the plastid genome is replaced by horizontal gene transfer. I suspect if the *ndh* genes were also introduced into the plastid genome of streptophytes from cyanobacteria independently from the acquisition of ribosomes. The plastid genome is not just a reduced endosymbiont genome, but has been reconstructed during the evolution of algae and plants.

8.5.5 Section Summary

In this section, I explained curious phylogenetic trees of plastid genes, which are possible traces of horizontal gene transfer in the plastid genome. The plastid genome is usually believed to result from the primary endosymbiosis, but the situation is not simple. There are two types of problems. First, some plastid-encoded proteins, such as NdhC, NdhE, RpoB, and RpoC1 + C2, present Type 2 phylogenetic trees. Type 2 trees were also found in red algal DnaB and DnaG encoded by the plastid genome. This could be an artifact of phylogenetic reconstruction, but some nuclear-encoded proteins also present Type 2 trees (Fig. 7.33). Type 2 trees imply that plastid genes originate from an ancestor of cyanobacteria. In the case of SQD2 in which the relative stem length is longer than the reference, we can imagine that SQD2 gene was a product of gene transfer from an ancestor of cyanobacteria to a proto-algal cell before chloroplast formation. If the relative stem length is not particularly longer or shorter than the reference, we have to assume a separate cyanobacteria-like lineage that kept the gene and transferred it later to an ancestor of algae. The cyanobacteria-like lineage might not give extant species, although I cannot present any exact image of such lineage.

The second problem is related to the exact position of plastid clade within the cyanobacterial clade. As described, plastid rRNA and ribosomal components have a deep origin in the cyanobacterial clade. They diverge after the basal group consisting of *Gloeobacter* and Yellowstone strains of *Synechococcus*, as well as *Pseudanabaena* (this was not included in most phylogenetic trees presented in this book). Some proteins, such as PsaA, PsaB, and MinD, diverged from the root of *Synechococcus-Prochlorococcus* group, called α-cyanobacteria. Some other proteins, such as RbcL (green lineage), RpoA, phosphoribulokinase (PRK), and ChlBLN, are related to β-cyanobacteria including *Synechocystis*, *Anabaena*, and *Cyanothece*. See Sato (2019) for individual phylogenetic trees. The branching point of Ndh proteins (as concatenated sequences) is also different. Although some of these could result from technical problems of phylogenetic reconstruction, we have to find a plausible solution. One idea is that plastids received multiple gene transfers from cyanobacteria. This is also related to the first problem. Normally, gene transfer to plastid genome is considered rare. But this assumes that exogenous genes come from bacteria other than cyanobacteria. The situation might be different if gene transfer from cyanobacteria is considered. How many times did cyanobacterial genomes contributed to the plastid genomes? The plastid genome might not be a stable entity, but was a product of gene transfers and, possibly, recombination. It is interesting to note that an objection to the minimizing or parsimony thinking on the endosymbiotic origins of chloroplasts was presented (Stiller 2014), but from a different viewpoint.

Chapter 9
Concluding Remarks

9.1 Knowledge Bias

Professor Patrick I. Keeling in the University of British Columbia, Canada, an active researcher on the secondary endosymbiosis, wrote an interesting opinion about the endosymbiotic theory of organelle origin (Keeling 2014), stating that a knowledge bias interferes with our thinking on the endosymbiotic theory and the origin of eukaryotic cells. This is a problem in the history of science, in which scientific interpretation is affected by preceding discoveries. He cited the schema notion proposed by a psychologist Jean Piaget. According to this theory, all data are interpreted according to the schema, and conflicting data are neglected. In other words, we have a cognitive inertia, in which accepted theory is hard to deny. In contrast, affirming accepted theory is not considered a novel scientific study. The presence of various conflicting proposals on the origin of eukaryotic cells is, according to Keeling, a result of inability of publishing papers supporting accepted theory. Three possibilities exist on the first branching group among eukaryotes, bacteria, and archaea, but all of them have inconsistencies, he affirmed, because all interpretations are affected by the initial impact of the contrast eukaryotes/prokaryotes.

If, however, all data are presented at a time, we might think in a different way, compared to the situation in which data are accumulated little by little. Currently, diversity in mitochondria and related organelles is known. If this fact were known at first, endosymbiotic theory of mitochondrial origin would not be justified. This is an interesting question of knowledge bias.

Keeling and McCutcheon (2017) stated that symbiosis is, in most cases, not mutualism but parasitism. If symbiosis is beneficial to both host and symbiont, then the symbiont will not be lost. Nevertheless, symbionts are often lost, and even organelles can be lost. In the genomic era, studies on endosymbiosis became interpretation of genomic data, affirming a lost organelle upon finding of a gene. Unfortunately, this cannot be demonstrated experimentally.

N. Sato, *Endosymbiotic Theories of Organelles Revisited*,
https://doi.org/10.1007/978-981-15-1161-5_9

After the death of Lynn Margulis in 2011, doubt about the endosymbiotic theory emerged and became seriously discussed. (Ab)use of endosymbiotic thinking tends to explain actual genomic complexity by a number of endosymbiosis. A question should be asked: what is the difference between endosymbiosis and gene transfer?

9.2 Change of Margulis and Paradigm Shift

In 2017, many articles on the endosymbiotic theory were published in *Journal of Theoretical Biology* to commemorate the 50th anniversary of the 1967 paper of Lynn Margulis. Some of them admire her works. Nevertheless, we have reasons to ask about her scientific achievement. As shown in Part I, we can hardly attribute new data or evidence to Margulis' contribution to the endosymbiotic theory, but Margulis assembled existing data to revitalize the endosymbiotic theory with her strong leadership. I present here my idea according to my previous article (Sato 2017). We have to recognize that many papers were published in the 1960s related to the endosymbiotic hypothesis on the origins of mitochondria and chloroplasts. Molecular phylogenetic studies beginning in the 1970s strongly suggested phylogenetic relationship between mitochondria and α-proteobacteria, as well as relationship between chloroplasts and cyanobacteria. Margulis was not directly involved in these sequencing and phylogenetic projects. Rather, her originality resided in the origin of mitosis in eukaryotes, as clearly stated in Sagan (1967) and Margulis (1970), namely, that centriole (or an endosymbiotic " 9 + 2" structure, supposedly a complex containing DNA) was the origin of centromere and kinetochore. Whether the endosymbiont was a spirochete or not was not a central problem in her initial writings. In her reasoning, she excluded chloroplasts from the main stream of eukaryotic evolution to deny "botanical myth." The endosymbiotic theory of chloroplast origin was nothing more than a necessary condition to propose her main scenario of eukaryotic origin. Nevertheless, her idea on the origin of mitosis was denied from the beginning, and only the endosymbiotic origin of mitochondria and chloroplasts was discussed in scientific community.

In 1967–1970, Lynn Margulis considered that the endosymbiotic origin of mitochondria and chloroplasts was evident owing to the works of other researchers. She was glad to learn that this theory was demonstrated by phylogenetic analysis in 1978 (see Lake 2011). However, her attitude seemed to change drastically later. She wrote in *Symbiotic Planet* in 1998:

> Symbiogenesis, an idea proposed by its Russian inventor Konstantin Merezhkovsky (1855–1921), refers to the formation of new organs and organisms through symbiotic mergers. As I will show it is a fundamental fact of evolution. All organisms large enough for us to see are composed of once independent microbes, teamed up to become larger wholes. As they merged, many lost what we in retrospect recognize as their former individuality. (p. 33).

> My best work, I believe, is the development of the details of the serial endosymbiosis theory [SET]. The central idea is that extra genes in the cytoplasm of animal, plant, and other nucleated cells are not "naked genes": rather they originated as bacterial genes. (p. 37).

> I am fond of bragging that we, my students and colleagues, have won three of the four battles of serial endosymbiosis theory (SET). We can now identify three of the four partners in the origin of cell individuality. (p. 38) [Note that the "four" included chloroplasts, mitochondria, flagella and basic eukaryotic cell. She considered that "the ground substance of cells, the nucleo-cytoplasm" descended from archaebacteria, or thermoacidophiles, and this was taken as "the third battle"].

Margulis mentioned here the name of Mereschkowsky (note that various English spelling exist for a Russian name) but did not show the exact reference. The readers would have an impression that Margulis and her team demonstrated entirely the endosymbiotic theory. We must admit that major evidence was provided by the cytological works in the 1960s and 1970s, and the phylogenetic analysis in the 1970s and 1980s. All these were achievements of other researchers. She mentioned her "best work," because, perhaps, her strong initiative on the endosymbiotic hypothesis prompted researchers to use phylogenetic analysis to demonstrate it.

We have to re-evaluate the endosymbiotic theory of Margulis. If all works of Margulis were totally erased from biology, we would have no serious problem in biology or history of biology. Nevertheless, we can affirm that a great achievement of Margulis would be her efforts in paradigm shift, namely, establishing the paradigm "endosymbiotic theory" in the 1980s (see Sect. 5.9.3). "Paradigm shift" is a term coined by Thomas Kuhn in his monumental work *The Structure of Scientific Revolution* (1962). Margulis cited Kuhn in her 1970 book (in Introduction) and her 1981 book (section "Paradigm Shift") to emphasize that it was a paradigm shift that she was promoting.

Originally, the endosymbiotic hypothesis was proposed by Mereschkowsky from 1905 onwards, and was known widely as "Famintzin-Mereschkowsky hypothesis" in the 1920s. However, it was not convincing to any biologist, because similarity of chloroplasts and cyanobacteria (blue-green algae) did not prove that chloroplasts were endosymbiotic cyanobacteria or their descendants. The hypothesis on the origin of organelles was not supported by experimental evidence. The situation seemed radically changed by the discovery of chloroplast DNA in the 1960s and comparative genomics of chloroplasts and cyanobacteria in later times.

In the 1970s and 1980s, biologists admired the great victory of molecular biology, believing that molecular biology was omnipotent in solving any biological problem. In developmental biology, in which molecular biology was not easily applied in early days, discovery of the Homeo box was a breakthrough in advancing molecular developmental biology. At that time, molecular biology laboratories were considered superior to other biology laboratories. Conventional taxonomy was replaced by molecular phylogenetics. "Gene" (nucleotide sequence rather than Mendelian factor) became a common language in all fields of biology. Under these conditions, the problem on the origin of organelles was replaced by the problem on the origin of organellar genomes. That must be the reality of the "paradigm shift" according to my interpretation, which is probably consistent with the concerns of Keeling (2014).

9.3 Rejecting the Simplified Scheme of Endosymbiosis in Textbooks

The endosymbiotic theory is now known to everyone (every biology students) and accepted as a fact, but little is known about its rather fragile evidence. Major evidence is the similarity (continuity) of chloroplasts and cyanobacteria, such as oxygenic photosynthesis, prokaryotic ribosome (protein synthesis), and prokaryotic RNA polymerase. As explained, similarity of membranes and presence of peptidoglycan are no longer evidence. In the early twentieth century, Mereschkowsky counted photosynthesis and protein synthesis as evidence for endosymbiotic origin of chloroplasts, although he did not know the notions of transcription and translation. The evidence seemed quite similar, at least formally. Although mitochondria are not a major topic in the present work, diversity in mitochondria and related organelles, such as hydrogenosomes and mitosomes, has been repeatedly emphasized by Bill Martin. It is not a simple story of endosymbiosis, as noted by Patrick I. Keeling. Historically, Kozo-Polyanski (1924) and Wallin (1927) stated endosymbiotic origin of mitochondria without solid evidence. Even in the era of genes and genomes, it remained unclear whether all mitochondria have a single origin, and whether endosymbiosis explained mitochondrial origin in a convincing way. We still have the same problem.

In the present work, I have been examining the endosymbiotic origin of chloroplasts/plastids, which I believe has more supporting evidence than the endosymbiotic origin of mitochondria. The theory did not change in its essence over the past 100 years. Comparison of homologous gene sequences seemed to provide strong, objective evidence, but it can only indicate phylogenetic relationship or ultimate origin. We have no means of explaining how a cyanobacterial protein (or a protein of other bacteria) came to exist in chloroplasts. I suppose that gene transfers occurred many times at different times. Various genes were introduced into proto-algal cells many times from cyanobacteria and other bacteria. Some of them remain encoded by the chloroplast genome, but many genes were now encoded by the nuclear genome. It is now time to change the simplistic textbook view of "primary endosymbiosis" as a single event occurring just once. The diagrammatic illustration of chloroplast formation by cyanobacterial endosymbiosis is also misleading, as the enzymes for the synthesis of membranes did not originate from cyanobacteria. It is a mistake of replacing an evolutionary event by a cytological event. Evolution is a dynamic process, in which a very rare mutation or event is fixed in a population as a result of genetic drift and natural selection. Biological evolution is different from the popularized notion of "evolution" describing social changes or "evolution" of Pokémons. Biological research has an important task to solve the complex dynamics of evolution.

9.4 Philosophical Reflections Added in English Version

I am closing the main text by some philosophical reflections on the logical structure of the endosymbiotic theory. The current, standard theory of endosymbiotic origin of chloroplasts is a hard-core theory. It consists of a three-step inference: First, we have phylogenetic data that chloroplast genome originated from cyanobacteria. Additionally, many chloroplast proteins originated from cyanobacteria. Second, we are forced to admit an evolutionary premise that evolutionary change must increase fitness. Therefore, acquisition of an entire organelle is supported by fitness increase, but individual components of an organelle do not contribute to gain fitness. Inference from these two arguments results in the notion that acquisition of a cyanobacterial endosymbiont as a whole created a chloroplast. This is a very strong argument that cannot be defeated by any new data. In this sense, this is a hard-core theory.

As I explained in the present work, we have many new data that do not fit this strong argument. All supporters of the theory tend to explain new data by ad hoc arguments to keep the core theory: these include later horizontal gene transfers, gene acquisition by endosymbiont before endosymbiosis, etc. But such a strong argument seems strange to me. Is it a hasty, impatient idea?

Facing new data, we may have two choices: either to give ad hoc explanations to non-cyanobacterial chloroplast proteins to keep the premise as a golden rule, or to relax the premise to reconcile it with new data. I think that the single premise is too strong in determining the conclusion of the inference. There are several ways to relax the premise or its usage: First, we may assume that various genes were introduced in small batches to assure functional merit at each step. Second, we may adopt neutral evolution rather than positive selection. A new gene may be fixed in the genome by chance. Third, we may suppose exaptation or preadaptation. A new gene had a certain function at first, but later gained a new function with another new gene. These are just examples of possible ideas. Evolution is an accumulation of very rare events. We might not be able to identify all processes occurring during the formation of chloroplasts. The idea of host-directed chloroplast formation that I explained in Sect. 8.4 is one of the possible evolutionary scenarios that use relaxed premises. I hope the readers will find a better explanatory model using a sober, relaxed premise.

Epilogue

The present work is a result of my cumulative researches, although my study still continues. It is perhaps rather a progress report summarizing the status of knowledge at this moment, from which we shall start. To write this volume, I am grateful to many teachers, research colleagues, and laboratory students.

My former PhD supervisor Dr. Norio Murata, and my former boss in the Faculty of Science, University of Tokyo, Dr. Tsuneyoshi Kuroiwa, gave me various knowledge and insights related to the endosymbiotic theory, although my researches of the time were not directly related to endosymbiosis. I remember the late Drs. Satoru Murakami, Shichiro Sato, and Hajime Ishikawa who taught me various aspects of the endosymbiotic theory. I also remember the late Dr. Ralph Lewin, who visited our laboratory and gave us an interesting seminar when I was a student. Dr. John Archibald in Dalhousie University, Canada, author of *One Plus One Equals One*, gave me invaluable additional information about the activities of Lynn Margulis. I obtained old literature from the perfect system of the library of the University of Tokyo. I wish to express my gratitude to my former colleagues: Dr. Hiroyoshi Takano (University of Kumamoto) for the study of peptidoglycan, Dr. Koichiro Awai (Shizuoka University) for the study of cyanobacterial lipids, Dr. Koichi Kobayashi (University of Tokyo, now Osaka Prefectural University) for the phylogenetic analysis of porphyrin biosynthesis enzymes. They helped me in starting this study of re-visiting the endosymbiotic theory. It was more than 10 years ago that I developed the comparative genomics software Gclust, which enabled comprehensive phylogenetic analysis in the present study. I constructed the Gclust server in 2009 with the help of a genomic project headed by Dr. Toshihisa Takagi (University of Tokyo). Gclust was not readily recognized as a bioinformatic tool, maybe because I was not originally in the informatics field, and similar software existed. Nevertheless, the phylogenetic analysis that I performed by using Gclust was successful and resulted in various research articles. Ten years ago, I did not recognize the close relationship of this software and the future study described in Chap. 7. Dr. Kenichiro Ishida (Tsukuba University) provided us with living cells of *Paulinella*, that we still continue to study. I also have to note that the late Dr. Terumitsu Hori

N. Sato, *Endosymbiotic Theories of Organelles Revisited*,
https://doi.org/10.1007/978-981-15-1161-5

(Toho University, and later Tsukuba University) taught me, 40 years ago, the technique of transmission electron microscopy which I used for taking some micrographs in the present work long after that. Members of my laboratory (closed in March 2019) helped me to take some photomicrographs in this book.

The writing of this book proceeded in parallel with writing research papers. My papers published in 2017 in the reference list were written, submitted, accepted, and published during the writing of the Japanese version of this book. The papers published later are also included in the reference list of the English version. The recent papers included those related to history of science, and phylogenetic analysis. Biology and science of history, both gave me the impetus of writing manuscripts. Biologists write rarely papers on history of science in their research field. Nevertheless, historical considerations can help scientific work. In this sense, I have been entirely devoted to the endosymbiotic theory since three years.

I am happy to recognize that all my efforts in the past 40 years are concentrated in the present work. However, this might be just a description of results, because we can only start a new activity based on the past achievements. The same is true for evolution. Organisms do what they can do. No external intension drives evolution of an organism. Perhaps many biologists are not satisfied with the statement that evolution is a result, but I suppose that everything is a result of something other. "Endosymbiotic" organelles are just results of successful cellular processes. No one had intentionally introduced a cyanobacterial cell into a eukaryotic host. If many gene transfers enabled the presence of chloroplasts, it is difficult to trace back the time to elucidate the history of formation of chloroplasts. In this sense, the studies presented in the current work is a small step towards the whole history of chloroplast formation. I hope my efforts will help the readers to gain new insights and start their study.

Many people helped me to realize this work. In addition to those I named already, I would like to thank Dr. Satoko Sekita (Kochi University) for supplying an illustration of coral. I thank Ms. Shiho Usuki (University of Tokyo Press) for editorial advice during the publication of the Japanese version, and Ms. Lee Mei Hann (Springer) for initial assistance in publication of the English version.

References[1]

Publications in English, German, French and Russian

Brightwell T (1858). Remarks on the genus "Rhizosolenia" of Ehrenberg. Quarterly Journal of Microscopical Science 6:93–95

Haeckel E (1866) *Generelle Morphologie der Organismen: allgemeine Grundzüge der organischen Formen-Wissenschaft, mechanisch begründet durch die von C Darwin reformirte Decendenz-Theorie.* Georg Reimer, Berlin

de Bary A (1879) *Die Erscheinung der Symbiose. Vortrag, gahalten auf der Versammlung deutscher Naturforscher und Aerzte zu Cassel.* Karl J Trübner, Strasburg

Schimper AFW (1883) Über die Entwicklung der Chlorophyllkörper und Farbkörper. Bot Z 41: 105–162

Schimper AFW (1885) Untersuchungen über die Chlorophyllkörner und die ihnen homologen Gebilde. Jahrb wiss Bot 16: 1–247

Altmann R (1889) Ueber die Fettumsetzungen im Organismus. Archiv für Anatomie und Physiologie Anatomischen Abtheilung. Supplement-Band. 86–104

Altmann R (1894) *Die Elementarorganismen und ihre Beziehungen zu den Zellen.* Zweite, vermehrte Auflage. Veit & Comp, Leipzig,

Lauterborn R (1895) Protozoenstudien. II. Paulinella chromatophora nov. gen. nov. spec., ein beschalter Rhizopode des Süßwassers mit blaugrünen chromatophorenartigen Einschlüssen. Z für wiss Zool 59: 537–544

Curie S (1899) Les rayons de Becquerl et le polonium. Revue générale des sciences pures et appliquées 10:41–50 [Note: S stands for Sklodowska, old family name of Marie Curie, that was erroneously used as given name]

Davis BM (1904) *Oltmanns' Morphologie und Biologie der Algen.* Erster Band. Spezieller Teil. Gustav Fischer, Jena

Mereschkowsky C (1905a) Über Natur und Ursprung der Chromatophoren im Pflanzenreiche. Biol Centralblatt 25:593–604

Mereschkowsky C (1905b) Nachtrag zu meiner Abhandlung: Über Nature und Ursprung der Chromatophoren im Pflanzenreiche. Biol Zentralblatt 25:689–701

[1] Journal articles and books are listed according to the year of publication. Within a year, they are sorted in the alphabetical order of the author(s). Note that the references described within the citations are not listed here.

N. Sato, *Endosymbiotic Theories of Organelles Revisited*,
https://doi.org/10.1007/978-981-15-1161-5

Famintzin A (1907) Die Symbiose als Mittel der Synthese von Organismen. Biol Zentralblatt 27:353–364

Mereschkowsky C (1910) Theorie der zwei Plasmaarten als Grundlage der Symbiogenesis, einer neuen Lehre von der Entstehung der Organismen. Biol Zentralblatt 30:278–288, 289-303,321-347,353-367

Minchin EA (1915) The evolution of the cell. Report of the Eighty-Fifth Meeting of the British Association for the Advancement of Science, September 7–11, 437–464. Reproduced in American Naturalist 50(1916), 3–38, 106–118, 271–283

Reinheimer H (1915) *Symbiogenesis. The Universal Law of Progressive Evolution*. Knapp Drewett and Sons, Westminster

Portier P (1918) *Les Symbiotes*. Masson, Paris

Mérejkovsky C (1920) La plante considérée comme un complexe symbiotique. Bull Soc Sci Nat Ouest France 6:17–98

Korschikov AA (1924) Pristologische Beobachtungen. Russ Archiv Pristolog 3:57–74

Kozo-Polyanski BM (1924/2010) *Symbiogenesis: A New Principle of Evolution*. Translated from Russian to English by Fet, V. Harvard University Press, Cambridge, MA. Russian original: Козо-Полянский Б. М. *Новый принцип биологии. Очерк теории симбиогенеза.*

Oparin AI (1924) The Origin of Life Translated by Synge, A. Dowloadable from http://www.uv.es/~orilife/englishindex.htm. Russian original: Опарин, А. И. *Происхождение жизни. (Proiskhozhdenie zhizni)*, Izd. Moskovskii Rabochi, Moscow

Wallin JE (1924) On the nature of mitochondria. VII. The independent growth of mitochondria in culture media. Amer J Anatomy 33:147–173

Wilson EB (1925) *The Cell in Development and Heredity*. 3rd edition. Macmillan, New York

Wallin JE (1927) *Symbionticism and the Origin of Species*. Williams and Wilkins, Baltimore

Haldane JBS (1929) The origin of life. *The Rationalist Annual* 3, 3–10 (page number is uncertain). Downloadable from http://www.uv.es/~orilife/englishindex.htm

Pascher A (1929a) Über die Natur der blaugrünen Chromatophoren des Rhizopoden Paulinella chromatophora. Zool Anz 81:189–194

Pascher A (1929b) Studien über Symbiosen. Über einige Endosymbiosen von Blaualgen in Einzellern Jahrb wiss Bot 71:386–462

Oparin AI (1936/1953) *The Origin of Life*. Translated by Morgulis, S., Dover, New York

Myer-Abich A (1950) Beiträge zur Theorie der Evolution der Organismen: II. Typensynthese durch Holobiose. Bibliotheca Biotheoretica 5:1–206. E. J. Brill, Leiden

Lederberg J (1952) Cell genetics and hereditary symbiosis. Physiol Rev 32:403–430

Buchner P (1953) *Endosymbiose der Tiere mit Pflanzlichen Mikroorganismen*. Birkhäuser, Basel. Revised English version: Endosymbiosis of animals with plant microorganisms. (1965) Translation by Bertha Nueller, with the collaboration of Francis H. Foeckler, Interscience

Stanier RY et al. (1957/1976) *The Microbial World.* Fourth edition. Prentice-Hall, Eglewood Cliffs, NJ

von Wettstein D (1959) Developmental changes in chloroplasts and their genetic control. In: Developmental Cytology. (Rudnick D ed) The Ronald Press, New York, pp. 123–160

Popper K (1959/2002) *The Logic of Scientific Discovery*. English edition Routledge, London

Stanier RY, van Niel CB (1962) The concept of a bacterium. Arch Mikrobiol 42:17–35

Ris H, Plaut W (1962) Ultrastructure of DNA-containing areas in the chloroplast of *Chlamydomonas*. J Cell Biol 13:383–391

Carson R (1962/2002) *Silent Spring*. Houghton Mifflin, Boston

Kuhn T (1962/1970) *The Structure of Scientific Revolutions*. Second Edition, Enlarged. University of Chicago Press, Chicago

Nass MMK, Nass S (1963a) Intramitochondrial fibers with DNA characteristics. I. fixation and electron staining reactions. J Cell Biol 19:583–611

Nass S, Nass MMK (1963b) Intramitochondrial fibers with DNA characteristics. II. Enzymatic and other hydrolytic treatments. J Cell Biol 19:613–629

Nutman PS, Moose B (eds) (1963) *Symbiotic Associations*. Thirteenth Symposium of the Society for General Microbiology held at the Royal Institution, London, April 1963. Cambridge University Press

Sager R, Ishida MR (1963) Chloroplast DNA in Chlamydomonas. Proc Natl Acad Sci USA 50:725–730

Cleveland LR, Grimstone AV (1964) The fine structure of the flagellate *Mixotricha paradoxa* and its associated micro-organisms. Proc R Soc B 159:668–686

Gibor A, Granick S (1964) Plastids and mitochondria: inheritable systems. Science 145:890–897

Echlin P, Morris I (1965) The relationship between blue-green algae and bacteria. Biol Rev 40:143–187

Edelman M et al (1965) Studies of chloroplast development in *Euglena*. XII. Two types of satellite DNA. J Mol Biol 11:769–774

Karakashian SJ, Siegel RW (1965), A genetic approach to endocellular symbiosis. Exp Parasitol 17:103–122

Nass S et al (1965a) Deoxyribonucleic acid in isolated rat-liver mitochondria. Biochim Biophys Acta 95:426–435

Nass MMK et al (1965b) The general occurrence of mitochondrial DNA. Exp Cell Res 37:516–539

Sagan L et al (1965) Studies on chloroplast development in *Euglena*. XI. Radioautographic localization of chloroplast DNA. Plant Physiol 40:1257–1260

Echlin P (1966) The cyanophytic origin of higher plant chloroplasts. Br phycol Bull 3:150–151

Haldar D et al (1966) Biogenesis of mitochondria. Nature 211:9–12

Bernal JD (1967) *The Origin of Life*. Weidenfeld and Nicolson, London

Edelman M et al (1967) Deoxyribonucleic acid of the blue-green algae (Cyanophyta). Bacteriol Rev 31:315–331

Fitch WM, Margoliash E (1967) Construction of phylogenetic trees. Science 155:279–284

Gibor A (1967) Inheritance of cytoplasmic organelles. In: Warren (1967) pp. 305–306

Goksøyr J (1967) Evolution of Eucaryotic cells. Nature 214:1161

Granick S, Gibor A (1967) The DNA of chloroplasts, mitochondria and centrioles. Prog Nucleic Acid Res Mol Biol 6:143–186

Kirk JTO, Tilney-Bassett RAE (1967) *The Plastids. Their chemistry, Structure, Growth and Inheritance.* Freeman, London

Klein R, Cronquist A (1967) A consideration of the evolutionary and taxonomic significance of some biochemical, micromorphological and physiological characteristics in the thallophytes. Quart Rev Biol 42:105–296

Sagan (Margulis) L (1967) On the origin of mitosing cells. J Theor Biol 14:225–274

Sager R (1967) Cytoplasmic genes and organelle formation. In: Warren. (1967) pp. 317–334

Warren KB (ed) (1967) *Formation and Fate of Cell Organelles*. Symposium of the International Society for Cell Biology. vol. 6, Academic Press, New York

Kimura M (1968) Evolutionary rate at the molecular level. Nature 217:624–626

Loening UE (1968) Molecular weights of ribosomal RNA in relation to evolution. J Mol Biol 38:355–365

Pickett-Heaps JD (1969) The evolution of the mitotic apparatus: an attempt at comparative ultrastructural cytology in dividing plant cells. Cytobios 3:257–280

Carr NG, Craig IW (1970) The relationships between bacteria, blue-green algae and chloroplasts. In: *Phytochemical Phylogeny* (Harborne JB, ed) Chapter 7, pp. 105–118, Academic Press, London

Cohen SS (1970) Are/Were mitochondria and chloroplasts microorganisms? Amer Scientists 58:281–289

Echlin P (1970) The origins of plants. In *Phytochemical Phylogeny* (Harborne JB, ed) Chapter 1, pp. 1–19, Academic Press, London

Goodenough UW, Levine RP (1970) The genetic activity of mitochondria and chloroplasts. Sci Amer 223:22–29

Margulis L (1970) *Origin of Eukaryotic Cells. Evidence and Research Implications for a Theory of the Origin and Evolution of Microbial, Plant, and Animal Cells on the Precambrian Earth.* Yale University Press, New Haven
Monod J (1970) *Le hasard et la nécessité.* Seuil, Paris
Nichols BW (1970) Comparative lipid biochemistry of photosynthetic organisms. In *Phytochemical Phylogeny* (Harborne JB, ed) Chapter 6, pp. 119–143, Academic Press, London
Raven PH (1970) A multiple origin for plastids and mitochondria. Science 169:641–646
Stanier RY (1970) Some aspects of the biology of cells and their possible evolutionary significance. Symp Soc Gen Microbiol 20:1–38
Baxter R (1971) Origin and continuity of mitochondria. In *Origin and Continuity of Cell Organelles* (Reinert J, Ursprung H, eds) pp. 46–64, Springer, Berlin
Fulton C (1971) Centrioles. In *Origin and Continuity of Cell Organelles* (Reinert J, Ursprung, H, eds) pp. 170–221, Springer, Berlin
Margulis L (1971) Symbiosis and Evolution. Sci Amer 225:48–57
Schnepf E, Brown RMJr (1971) On relationships between endosymbiosis and the origin of plastids and mitochondria. In *Origin and Continuity of Cell Organelles* (Reinert J, Ursprung H, eds) pp. 299–332, Springer, Berlin
Stubbe W (1971) Origin and continuity of plastids. In *Origin and Continuity of Cell Organelles*, (Reinert J, Ursprung H, eds) pp. 65–81, Springer, Berlin
Hall DO, Rao KK (1972/1999) *Photosynthesis.* Sixth edition. Cambridge University Press, Cambridge
Lee RE (1972) Origin of plastids and the phylogeny of algae. Nature 237: 44–46
Raff RA, Mahler HR (1972) The non symbiotic origin of mitochondria. Science 177:575–582
Cohen SS (1973) Mitochondria and chloroplasts revisited. Amer Scientists 61:437–445
Takhtajan AL (1973) Four kingdoms of the living world [in Russian]. *Priroda* (Nature) 2:22–32. Russian original: А. Л. Тахтаджян: Четыре царства органического мира. природа.
Dayhoff MO et al (1974) Inferences from protein and nucleic acid sequences: early molecular evolution, divergence of kingdoms and rates of change. Orig Life 5:311–330
Stanier RY (1974) The origins of photosynthesis in eukaryotes. Symp Soc Gen Microbiol 24:219–242
Taylor DL (1974a) A multiple origin for plastids and mitochondria. Science 170:1332
Taylor FJR (1974b) Implications and extensions of the serial endosymbiosis theory of the origin of eukaryotes. Taxon 23:229–258
Uzzell T, Spolsky C (1974) Mitochondria and plastids as endosymbionts: a rivival of special creation? The similarities between cellular organelles and prokaryotes are probably primitive features retained independently from a common ancestor. Am Sci 62:334–343
Bogorad L (1975) Evolution of organelles and eukaryotic genomes. Science 188:891–898
Bonen L, Doolittle WF (1975) On the prokaryotic nature of the red algal chloroplasts. Proc Natl Acad Sci USA 72:2310–2314
Cavalier-Smith T (1975) The origin of nuclei and eukaryotic cells. Nature 256:463–468
Lewin RA, Withers NW (1975) Extraordinary pigment composition of a prokaryotic alga. Nature 256:735–737
Margulis L (1975) Symbiotic theory of the origin of eukaryotic organelles: criteria for proof. Symp Soc Exp Biol 29:21–38
Newcomb NH, Pugh TD (1975) Blue-green algae associated with ascidians of the Great Barrier Reef. Nature 253:533–534
Zablen LB et al (1975) Phylogenetic origin of the chloroplast and prokaryotic nature of its ribosomal RNA. Proc Natl Acad Sci USA 72:2418–2422
Bonen L, Doolittle WF (1976) Partial sequences of 16S rRNA and the phylogeny of blue-green algae and chloroplasts. Nature 261:669–673
Buetow DE (1976) Phylogenetic origin of chloroplast. J Protozool 23:41–47
Bonen L et al (1977) Wheat embryo mitochondrial 18S ribosomal RNA: evidence for its prokaryotic nature. Nucleic Acids Res 4:663–671

Ishikawa H (1977) Evolution of ribosomal RNA. Comp Biochem Physiol B 58:1–7
Schwartz RM, Dayhoff MO (1978) Origins of prokaryotes, eukaryotes, mitochondria, and chloroplasts. A perspective is derived from protein and nucleic acid sequence data. Science 199:395–403
Khakhina LN (1979/1992) *Concepts of Symbiogenesis. A Historical and Critical Study of the Research of Russian Botanists.* Edited by Margulis L, McMenamin M, Translated by Merkel S, Coalson R. Yale University Press, New Haven. Russian original: Хахина Л.Н. Проблема симбиогенеза // Историко-критический очерк исследовании отечественных ботаников – Л.: Наука (1979)
Taylor FJR (1979) Symbionticism revisited: a discussion of the evolutionary impact of intracellular symbioses. Proc R Soc Lond B 204:267–286
Feige G et al (1980) Discovery of a new glyceroglycolipid in blue-green algae and its role in galactolipid biosynthesis. In *Biogenesis and Function of Plant Lipids*, Mazliak P, Benveniste P, Costes C, Douce R (eds), Elsevier/NorthHolland Biomedical Press, Amsterdam. pp. 135–140
Lee RE (1980) *Phycology*. Cambridge University Press, Cambridge
Doolittle WF, Bonen L (1981) Molecular sequence data indicating an endosymbiotic origin for plastids. Annals N Y Acad Sci 361:248–259
Margulis L (1981/1993) *Symbiosis in Cell Evolution*. 2nd Ed. *Microbial Communities in the Archean and Proteozoic Eons*. Freeman, San Francisco
Whatley JM (1981) Chloroplast evolution—ancient and modern. Ann N Y Acad Sci 361:154–164
Lewin RA (1981) *Prochloron* and the theory of symbiogenesis. Ann N Y Acad Sci 361:325–329
Sato N, Murata N (1982) Lipid biosynthesis in the blue-green alga, *Anabaena variabilis*. I. Lipid classes. Biochim Biophys Acta 710:271–278
Gray MW, Doolittle WF (1982) Has the endosymbiont hypothesis been proven? Microbiol Rev 46:1–42
Wallace DC (1982) Structure and evolution of organelle genomes. Microbiol Rev 46:208–240
Cavalier-Smith T (1982) The origin of plastids. Biol J Linn Soc 17:289–306
Murata N, Sato N (1983) Analysis of lipids in *Prochloron* sp.: Occurrence of monoglucosyl diacylglycerol. Plant Cell Physiol 24:133–138
Margulis L, Bermudes D (1985) Symbiosis as a mechanism of evolution: status of cell symbiosis theory. Symbiosis 1:101–124
Cavalier-Smith T (1987) The simultaneous symbiotic origin of mitochondria, chloroplasts, and microbodies. Annals N Y Acad Sci 503:55–71
Taylor FJR (1987) An overview of the status of evolutionary cell symbiosis theories. Annals N Y Acad Sci 503:1–16
Hall JL et al (1989) Basal body/centriolar DNA: Molecular genetic studies in Chlamydomonas. Cell 59:121–132
Johnson KA, Rosenbaum JL (1991) Basal bodies and DNA. Trends Cell Sci 1:145–149
Gray MW (1992) The endosymbiont hypothesis revisited. Int Rev Cytol 141:233–357
Palenik R, Haselkorn R (1992) Multiple evolutionary origins of prochlorophytes, the chlorophyll *b*-containing prokaryotes. Nature 355:265–267
Urbach E et al (1992) Multiple evolutionary origins of prochlorophytes within the cyanobacterial radiation. Nature 355:267–270
Sapp J (1994) *Evolution by Association*. Oxford University Press, Oxford
McFadden G, Gilson P (1995) Something borrowed, something green: lateral transfer of chloroplasts by secondary endosymbiosis. Trends Evol Ecol 10:12–17
Hess WR et al (1996) Coexistence of phycoerythrin and a chlorophyll *a*/*b* antenna in a marine prokaryote. Proc Natl Acad Sci USA 93:1126–1130
Brinkley W (1997) Microtubules: a brief historical perspective. J Struct Biol 118:84–86
Margulis L (1998) *Symbiotic Planet. A New Look at Evolution*. Basic Books, Amherst, MA
Sicheritz-Pontén T et al (1998) A phylogenetic analysis of the cytochrome *b* and cytochrome *c* oxidase I genes supports an origin of mitochondria from within the Rickettsiaceae. Biochim Biophys Acta 1365:545–551

Martin W, Kowallik K (1999) Annotated English translation of Mereschkowsky's 1905 paper 'Über Natur und Ursprung der Chromatophoren im Pflanzenreiche'. Eur J Phycol 34:287–295

Tabita FR (1999) Microbial ribulose 1,5-bisphospate carboxylase/oxygenase: A different perspective. Photosynth Res 60:1–28

Cavalier-Smith T (2000) Membrane heredity and early chloroplast evolution. Trends Plant Sci 5:174–182

Karlberg O et al (2000) The dual origin of the yeast mitochondrial proteome. Yeast 17:170–187

Moreira D et al (2000) The origin of red algae and the evolution of chloroplasts. Nature 405:69–72

Sato N (2001) Was the evolution of plastid genetic machinery discontinuous? Trends Plant Sci 6:151–156

Douglas S et al (2001) The highly reduced genome of an enslaved algal nucleus. Nature 410:1091–1096

Buchanan BB et al (eds) (2002) *Biochemistry and Molecular Biology of Plants*. American Society of Plant Physiologists, Rockville, Maryland

Cain M et al (2002/2012) *Discover Biology*. Second/Fifth Edition. W. W. Norton & Co. Inc., New York, NY

Martin W et al (2002) Evolutionary analysis of Arabidopsis, cyanobacterial, and chloroplast genomes reveals plastid phylogeny and thousands of cyanobacterial genes in the nucleus. Proc Natl Acad Sci USA 99:12246–12251

Sapp J et al (2002) Symbiogenesis: The hidden face of Constantin Merezhkowsky. Hist Philos Life Sci 24:413–440

Tielens AGM et al (2002) Mitochondria as we don't know them. Trends Biochem Sci 27:564–572

Dufresne A et al (2003) Genome sequence of the cyanobacterium *Prochlorococcus marinus* SS120, a nearly minimal oxyphototrophic genome. Proc Natl Acad Sci USA 100:10020–10025

Rocap G et al (2003) Genome divergence in two *Prochlorococcus* ecotypes reflects oceanic niche differentiation. Nature 424:1042–1047

Matsuzaki M et al (2004) Genome sequence of the ultrasmall unicellular red alga *Cyanidioschyzon merolae* 10D. Nature 428:653–657

Adl SM et al (2005) The new higher level classification of Eukaryotes with emphasis on the taxonomy of protists. J Eukaryot Microbiol 52:399–451

Jenke-Kodama H et al (2005) Evolutionary implications of bacterial polyketide synthases. Mol Biol Evol 22:2027–2039

Miyagishima S (2005) Origin and evolution of the chloroplast division machinery. J Plant Res 118:295–306

Sato N et al (2005) Mass identification of chloroplast proteins of endosymbiont origin by phylogenetic profiling based on organism-optimized homologous protein groups. Genome Inform 16:56–68

Fitzpatrick DA et al (2006) Genome phylogenies indicate a meaningful α-proteobacterial phylogeny and support a grouping of the mitochondria with the Rickettsiales. Mol Biol Evol 23:74–85

Okamoto N, Inoue I (2006) *Hatena arenicola* gen. et sp. nov., a katablepharid undergoing probable plastid acquisition. Protist 157:401–419

Sato N (2006) Origin and evolution of plastids: genomic view on the unification and diversity of plastids. In: Wise RR, Hoober JK (eds), *The Structure and Function of Plastids*, Chapter 4, pp. 75–102. Springer, Berlin

Martin WF, Müller M (eds) (2007) *Origin of Mitochondria and Hydrogenosomes*. Springer, Berlin

Sato N, Moriyama T (2007) Genomic and biochemical analysis of lipid biosynthesis in the unicellular rhodophyte *Cyanidioschyzon merolae*: lack of plastidic desaturation pathway results in mixed pathway of galactolipid synthesis. Eukaryot Cell 6:1006–1017

Gould SB et al (2008) Plastid evolution. Annu Rev Plant Biol 59:491–517

Harris E (ed) (2008) *The Chlamydomonas Sourcebook*, 2nd Edition. Academic Press, Amsterdam

Nowack ECM et al (2008) Chromatophore genome sequence of *Paulinella* sheds light on acquisition of photosynthesis by eukaryotes. Curr. Biol 18:410–418

Zehr JP et al (2008) Globally distributed uncultivated oceanic N_2-fixing cyanobacteria lack oxygenic photosystem II. Science 322:1110–1112
Benchimol M (2009) Hydrogenosomes under microscopy. Tissue and Cell 41:151-168*trans*
Sato N (2009) Gclust: *trans*-kingdom classification of proteins using automatic individual threshold setting. Bioinformatics 25:599–605
Maier T et al (2010) Structure and function of eukaryotic fatty acid synthases. Quart Rev Biophys 43:373–472
Malaterre C (2010) *Les origines de la vie. Émergence ou explication réductive*? Hermann, Paris
Taiz L, Zeiger E (2010) *Plant Physiology*. Fifth Edition. Sinauer Associates, Sunderland, MA
Takano H, Takechi K (2010) Plastid peptidoglycan. Biochim Biophys Acta 1800:144–151
Lake JA (2011) Lynn Margulis (1938–2011). Nature 480:458
Shigenobu A, Wilson ACC (2011) Genomic revelations of a mutualism: the pea aphid and its obligate bacterial symbiont. Cell Mol Life Sci 68:1297–1309
Reinhold-Hurek B, Hurek T (2011) Living inside plants: bacterial endophytes. Curr Opin Plant Biol 14:435–443
Lang BF, Burger G (2012) Mitochondrial and eukaryotic origins: A critical review. Adv Bot Res 63:1–19
Liu Y, Bassham DC (2012) Autophagy: Pathways for self-eating in plant cells. Annu Rev Plant Biol 63:215–237
Price DC et al (2012) *Cyanophora paradoxa* genome elucidates origin of photosynthesis in algae and plants. Science 335:843–847
Sagan D (ed) (2012) *Lynn Margulis. The Life and Legacy of a Scientific Rebel*. Chelsea Green Publishing, White River Junction, VT
Sapp J (2012) Too fantastic for polite society: A brief history of symbiosis theory. In: Sagan D (2012), pp. 54–67
Sato N (2012) Scientific élan vital: Entropy deficit or inhomogeneity as a unified concept of driving forces of life in hierarchical biosphere driven by photosynthesis. Entropy 14:233–251
Qiu H et al (2012) Ancient gene paralogy may mislead inference of plastid phylogeny. Mol Biol Evol 29:3333–3343
Bodył A, Mackiewicz P (2013) Endosymbiotic Theory. *Brenner's Encyclopedia of Genetics* (Second Edition), 2013, pp. 484–492
Burger G et al (2013) Strikingly bacteria-like and gene-rich mitochondrial genomes throughout jakobid protists. Genome Biol Evol 5:418–438
Hagino K et al (2013) Discovery of an endosymbiotic nitrogen-fixing cyanobacterium UCYN-A in *Braarudosphaera bigelowii* (Prymnesiophyceae). PLoS One 8:e81749
Harvath SE, Daum G (2013) Lipids of mitochondria. Prog Lipid Res 52:590–614
Qiu H et al (2013) Assessing the bacterial contribution to the plastid proteome. Trends Plant Sci 18:680–687
Shih PM et al (2013) Improving the coverage of the cyanobacterial phylum using diversity-driven genome sequencing. Proc Natl Acad Sci USA 110:1053–1058
Archibald J (2014) *One Plus One Equals One. Symbiosis and the evolution of complex life*. Oxford University Press, Oxford
Bailly X et al (2014) The chimerical and multifaceted marine acoel *Symsagittifera roscoffensis*: from photosymbiosis to brain regeneration. Front Microbiol 5:498
Campbell NA et al (2014/2016) *Biology*. 10th Edition [11th Edition is the current version], Pearson Education, San Francisco.
Gray MW (2014) The pre-endosymbiont hypothesis: A new perspective on the origin and evolution of mitochondria. Cold Spring Harb Perspect Biol 6:a016097
He D et al (2014) An alternative root for the eukaryote tree of life. Curr Biol 24:465–470
Keeling PJ (2014) The impact of history on our perception of evolutionary events: endosymbiosis and the origin of eukaryotic complexity. Cold Spring Harb Perspect Biol 6: a016196

Kobayashi K et al (2014) Molecular phylogeny and intricate evolutionary history of the three isofunctional enzymes involved in the oxidation of Protoporphyrinogen IX. Genome Biol Evol 6:2141–2155

Löffelhardt W (ed) (2014) *Endosymbiosis*. Springer, Wien

Moriyama T, Sato N (2014) Enzymes involved in organellar DNA replication in photosynthetic eukaryotes. Front Plant Sci 5:480

Moriyama T et al (2014) Localization and phylogenetic analysis of enzymes related to organellar genome replication in the unicellular rhodophyte *Cyanidioschyzon merolae*. Genome Biol Evol 6:228–237

Nakayama T et al (2014) Complete genome of a nonphotosynthetic cyanobacterium in a diatom reveals recent adaptations to an intracellular lifestyle. Proc Natl Acad Sci USA 111:11407–11412

Stiller JW (2014) Toward an empirical framework for interpreting plastid evolution. J Phycol 50:462–471

Yamaguchi H et al (2014) Molecular diversity of endosymbiotic *Nephroselmis* (Nephroselmidophyceae) in *Hatena arenicola* (Katablepharidophycota). J Plant Res 127:241–247

Archibald JM (2015) Endosymbiosis and eukaryotic cell evolution. Curr Biol 25:R911–R921

Ball SG et al (2015) Toward an understanding of the function of Chlamydiales in plastid endosymbiosis. Biochim Biophys Acta 1847:495–504

Börner T et al (2015) Chloroplast RNA-polymerases: Role in chloroplast biogenesis. Biochim Biophys Acta 1847:761–769

Fu J et al (2015) The centrosome and its duplication cycle. Cold Spring Harb Perspect Biol 7:a015800

Gray MW (2015) Mosaic nature of the mitochondrial proteome: Implications for the origin and evolution of mitochondria. Proc Natl Acad Sci USA 112:10133–10138

Ku C et al (2015) Endosymbiotic gene transfer from prokaryotic pangenomes: Inherited chimerism in eukaryotes. Proc Natl Acad Sci USA 112:10139–10146

Lin S et al (2015) The *Symbiodinium kawagutii* genome illuminates dinoflagellate gene expression and coral symbiosis. Science 350:691–694

Martin WF et al (2015) Endosymbiotic theories for eukaryote origin. Philos Trans R Soc Lond B Biol Sci 370:20140330

Paila YD et al (2015) New insights into the mechanism of chloroplast protein import and its integration with protein quality control, organelle biogenesis and development. J Mol Biol 427:1038–1060

Rast A et al (2015) Biogenesis of thylakoid membranes. Biochim Biophys Acta 1847: 821–830

Spang A et al (2015) Complex archaea that bridge the gap between prokaryotes and eukaryotes. Nature 521:173–179

Sato N (2015) Is monoglucosyl diacylglycerol a precursor to monogalactosyl diacylglycerol in all cyanobacteria? Plant Cell Physiol 56:1890–1899

Zhang L, Sakamoto W (2015) Possible function of VIPP1 in mainaining chloroplast membranes. Biochim Biophys Acta 1847: 831–837

Ahmed T et al (2016) Cryo-EM structure of the large subunit of the spinach chloroplast ribosome. Sci Rep 6:35793

Aranda M et al (2016) Genomes of coral dinoflaggellate symbionts highlight evolutionary adaptations conductive to a symbiotic lifestyle. Sci Rep 6:39734

Boisy G et al (2016) Microtubules: 50 years on from the discovery of tubulin. Nature Review Mol Cell Biol 17:322–328

Caforio A, Driessen AJM (2016) Archaeal phospholipids: Structural properties and biosynthesis. Biochim Biophys Acta 1862:1325–1339

Gould SB et al (2016) Bacterial vesicle secretion and the evolutionary origin of the eukaryotic endomembrane system. Trends Microbiol 24:525–534

Morange M (2016) *Une histoire de la biologie*, Seuil, Paris

Nowack ECM et al (2016) Gene transfers from diverse bacteria compensate for reductive genome evolution in the chromatophore of *Paulinella chromatophora*. Proc Natl Acad Sci USA 113:12214–12219

Pittis AA, Gabaldón T (2016) Late acquisition of mitochondria by a host with chimaeric prokaryotic ancestry. Nature 531:101–104

Sato N, Awai K (2016) Diversity in biosynthetic pathways of galactolipids in the light of endosymbiotic origin of chloroplasts. Front Plant Sci 7:117

Sato N et al (2016) Diverse pathways of biosynthesis of phosphatidylcholine in algae as estimated by labeling studies and genomic sequence analysis. Plant J 87:281–292

Shikanai T (2016) Chloroplast NDH: a different enzyme with a structure similar to that of respiratory NADH dehydrogenase. Biochim Biophys Acta 1857:1015–1022

Tajima N et al (2016) Sequencing and analysis of the complete organellar genomes of Parmales, a closely related group to Bacillariophyta (diatoms). Curr Genet 62:887–896

Driscoll TP et al. (2017) Wholly Rickettsia! Reconstructed metabolic profile of the quintessential bacterial parasite of eukaryotic cells. mBio 26:e00859–17

Grosche C, Rensing, SA (2017) Three rings for the evolution of plastid shape: a tale of land plant FtsZ. Protoplasma 254:1879–1885

Harish A, Kurland CG (2017) Mitochondria are not captive bacteria. J Theor Biol 434:88–98

Keeling PJ, McCutcheon JP (2017) Endosymbiosis: the feeling is not mutual. J Theor Biol 434:75–79

Lane N (2017) Serial endosymbiosis or singular event at the origin of eukaryotes? J Theor Biol 434:58–67

Lazcano A, Peretó J (2017) On the origin of mitosing cells: a historical appraisal of Lynn Margulis endosymbiotic theory. J Theor Biol 434:80–87

Lhee D et al (2017) Diversity of the photosynthetic *Paulinella* species, with the description of *Paulinella micropora* sp. nov. and the chromatophore genome sequence for strain KR01. Protist 168:155–170

López-García P et al (2017) Symbiosis in eukaryotic evolution. J Theor Biol 434:20–33

Luginbuehl LH et al (2017) Fatty acids in arbuscular mycorrhizal fungi are synthesized by the host plant. Science 356:1175–1178

Martin WF (2017) Physiology, anaerobes, and the origin of mitosing cells 50 years on. J Theor Biol 434:2–10

Ponse-Toledo RI et al (2017) An early-branching freshwater cyanobacterium at the origin of plastids. Curr Biol 27:386–391

Roger AJ et al (2017) The origin and diversification of mitochondria. Curr Biol 27:R1177–R1192

Sato N (2017) Revisiting theoretical basis of the endosymbiotic origin of plastids in the original context of Lynn Margulis on the origin of mitosing, eukaryotic cells. J Theor Biol 434:104–113

Sato N, Takano H (2017) Diverse origins of enzymes involved in the biosynthesis of chloroplast peptidoglycan. J Plant Res 130:635–645

Sato N et al (2017) Single-pixel densitometry revealed the presence of peptidoglycan in the intermembrane space of moss chloroplast envelope in conventional electron micrographs. Plant Cell Physiol 58:1743–1751

Thornhill DJ et al (2017) Population genetics of reef coral endosymbionts (*Symbiodinium*, Dinophyceae). Mol Ecol 26:2640–2659

Tria FDK et al (2017) Phylogenetic rooting using minimal ancestor deviation. Nat Ecol Evol 1:0193

Sato N, Awai K (2017) "Prokaryotic Pathway" is not prokaryotic: noncyanobacterial origin of the chloroplast lipid biosynthetic pathway revealed by comprehensive phylogenomic analysis. Genome Biol Evol 9:3162–3178

Hirashima T et al (2018) Evolution of the phosphatidylcholine biosynthesis pathways in green algae: combinatorial diversity of methyltransferases. J Mol Evol 86:68–76

Maréchal E (2018) Primary Endosymbiosis: Emergence of the Primary Chloroplast and the Chromatophore, Two Independent Events. Chapter 1 *In* Maréchal E (ed) *Plastids: Methods and Protocols*. Methods in Molecular Biology, 1829, Humana Press, New York, pp. 3–16

Martijn J et al (2018) Deep mitochondrial origin outside the sampled alphaproteobacteria. Nature 557:101–105

Adl SM et al (2019) Revisions to the classification, nomenclature, and diversity of Eukaryotes. J Eukaryotic Microbiol 66:4-119

Schuller JM et al (2019) Structural adaptations of photosynthetic complex I enable ferredoxin-dependent electron transfer. Science 363:257–260

Laughlin TG et al (2019) Structure of the complex I-like molecule NDH of oxygenic photosynthesis. Nature 566:411–414

Sato N (2019) Complex origins of chloroplast membranes with photosynthetic machineries: Multiple transfers of genes from divergent organisms at different times or a single endosymbiotic event? J Plant Res (Open Access) https://doi.org/10.1007/s10265-019-01157-z.

Publications in Japanese[2]

Furuya M et al. (eds) (1971J) *Photosynthesis*. Asakura Shoten, Tokyo. No direct mention of endosymbiotic theory was provided, but a short comment was added by Taniguchi, S. in proof: “A general discussion on probable evolutionary relationship between prokaryotic and eukaryotic cells through development of photosynthetic systems” p. 181. My translation

Kato S (1973J) *Introduction to Photosynthesis.* Kyoritsu Shuppan, Tokyo. Importance of photosynthesis in the Earth history was stated, but no description was provided on algae, cyanobacteria, or endosymbiotic theory

Sato S (1975/1979J) *The Cell.* UP BIOLOGY, Second Edition, University of Tokyo Press. In Chapter 9 Origin of Eukaryotic Cells, the author’s hypothesis on the origin of eukaryotic cell was illustrated by citing Mereschkowsky and Margulis. He wrote: “Margulis (Lynn Sagan in 1967) proposed ambitiously an endosymbiotic theory in her paper ‘On the origin of mitosing cells’ based on recent biochemical studies. In 1970, she systematically formulated this theory with geochemical data, and published *Origin of Eukaryotic Cells*, which drew attention of many scientists. Then, this hypothesis was examined by various fields, such as biochemistry, microbiology and phylogeny. (…) Margulis’ theory contained some obvious errors, namely, the confusion that she considered eukaryotic and prokaryotic flagella homologous, and the belief that basal body contained DNA. Here I will present the endosymbiotic theory after removing these errors and reconstructing according to my view.” (p. 114, my translation)

Kagawa Y (1975J) *Biological Membranes and Biological Energy.* UPBIOLOGY, University of Tokyo Press. The author commented on the endosymbiotic theory as follows: “Both chloroplasts and mitochondria have their own DNA, proliferate autonomously, and mediate cytoplasmic inheritance, which are parasitic characters. In fact, aerobic bacteria, which are prokaryotic, are very similar to mitochondria. It is highly probable that they enter a eukaryotic cell to become mitochondria in ancient times. (p. 6, my translation)

Nakamura H (1987J) Origin of mitochondrion and chloroplast in the Eukaryotic cell. Tanpakushitsu Kakusan Kouso (PNE) 32:1019–1030. The author presented arguments against the endosymbiotic theory

Sato S (1988J) *Evolution of Cells.* University of Tokyo Press. Various theories on the origin of organelles are presented, including the work of a Japanese biologist, Shozaburo Watase. The

[2] Japanese references are cited in the main text with “J” after the year, such as Kato (1973J). The title of each article was translated by the author, if English title is not provided in the original publications. A short comment or explanation is added for each publication, if necessary.

author commented on the priority of the endosymbiotic theory: "As I mentioned above, we cannot say that Margulis was a pioneer that revitalized the endosymbiotic theory. Nevertheless, we have an impression as if she proposed the modern endosymbiotic theory. This could be due to her extraordinary activity." (p. 20, my translation)

Watanabe I (1992J) Symbiosis of *Azolla* with cyanobacteria. Kagaku to Seibutsu (Chemistry and Biology) 30:820–829

Kuroiwa T (2000J) *Origin of Mitochondria. History of Four Billion Years*. NHK Books, Tokyo

Sato N (2002J) Diversity in molecular organization and function of plastid nucleoids – a hypothesis on the discontinuous evolution of plastid genomic machinery – Seikagaku Journal of Japanese Biochemical Society 74:27–37

Photosynthesis Education Team, The University of Tokyo (2007J) *The Science of Photosynthesis*. University of Tokyo Press, Tokyo

Inoue I (2006/2007J) *The Natural History of Algae*. Second edition. Tokai University Press, Tokyo

Kitade O (2007J) Characteristics and host-symbiont relationships of termite gut flagellates. Jap J Protozool 40:101–112

Yamamoto Y (2008J) Photosynthetic sea slugs having kleptoplasts Newsletter of The Japanese Society of Photosynthesis Research 18:42–45

Editorial Committee for the Biology Illustrated, College of Arts and Sciences, The University of Tokyo (ed) (2010J) *Biology Illustrated*. University of Tokyo Press, Tokyo

Sato N (2012J) *40 Years After "Chance and Necessity". Reconstruction of Monod's view on life, evolution, and the human future*. University of Tokyo Press, Tokyo

Ishida H (2014J) Plant nutrition recycling and chloroplast autophagy. Kagaku to Seibutsu (Chemistry and Biology) 52:610–615

Sato N (2014J) *Plant Physiology, basic mechanisms and principles*. Shokabo, Tokyo

Hayakawa M, Suzaki T (2016J) Endosymbiosis in *Paramecium bursaria*. Hikaku Seiri Seikagaku Comparative Physiology and Biochemistry 33:108–115

Yoshimoto K (2014J) Significance and roles of autophagy in plants. Kagaku to Seibutsu (Chemistry and Biology) 52:535–540

Sato N (2015J) Glycolipid biosynthesis and oxygenic photosynthesis in cyanobacteria. Seikagaku Journal of Japanese Biochemical Society 87:209–211

Kawaguchi M (2016J) Long-distance regulation of root nodule formation by leaf-root communications. Kagaku to Seibutsu (Chemistry and Biology) 54:94–101

Sato N (2016J) Origins of endosymbiotic theory of plastids: Japanese translation of the Mereschkowsky's paper in 1905 with introductory notes. Newsletter of the Japanese Society of Photosynthesis Research 26:106–117

Kitagawa D (2017J) Conserved molecular mechanisms of centrosome duplication and maturation. Seikagaku Journal of Japanese Biochemical Society 89:489–497

MIX
Papier aus verantwortungsvollen Quellen
Paper from responsible sources
FSC® C105338

If you have any concerns about our products,
you can contact us on
ProductSafety@springernature.com

In case Publisher is established outside the EU,
the EU authorized representative is:
Springer Nature Customer Service Center GmbH
Europaplatz 3, 69115 Heidelberg, Germany

Printed by Libri Plureos GmbH
in Hamburg, Germany